BAYESIAN MODELING OF UNCERTAINTY
IN LOW-LEVEL VISION

T0332538

THE KLUWER INTERNATIONAL SERIES IN
ENGINEERING AND COMPUTER SCIENCE

ROBOTICS: VISION, MANIPULATION AND SENSORS

Consulting Editor

Takeo Kanade
Carnegie Mellon University

BAYESIAN MODELING OF UNCERTAINTY IN LOW-LEVEL VISION

by

Richard Szeliski
Carnegie Mellon University

with a foreword by
Takeo Kanade

KLUWER ACADEMIC PUBLISHERS
Boston/Dordrecht/London

Distributors for North America:
Kluwer Academic Publishers
101 Philip Drive
Assinippi Park
Norwell, Massachusetts 02061 USA

Distributors for all other countries:
Kluwer Academic Publishers Group
Distribution Centre
Post Office Box 322
3300 AH Dordrecht, THE NETHERLANDS

Library of Congress Cataloging-in-Publication Data

Szeliski, Richard. 1958–
  Bayesian modeling of uncertainty in low-level vision / Richard
Szeliski ; foreword by Takeo Kanade.
    p. cm. — (The Kluwer international series in engineering and
computer science ; #79)
    ISBN 0-7923-9039-3
    1. Computer vision—Mathematical models. I. Title. II. Series:
Kluwer international series in engineering and computer science :
SECS 79.
TA1632.S94 1989   89–15632
                                                        CIP

Printed in the United States of America

To my parents, Zdzisław Ludwig Szeliski and Jadwiga Halina Szeliska, who have always encouraged my academic aspirations, and whose integrity and love have been a lifelong inspiration.

Contents

List of Figures

Foreword

Vision has to deal with uncertainty. The sensors are noisy, the prior knowledge is uncertain or inaccurate, and the problems of recovering scene information from images are often ill-posed or underconstrained. This research monograph, which is based on Richard Szeliski's Ph.D. dissertation at Carnegie Mellon University, presents a Bayesian model for representing and processing uncertainty in low-level vision.

Recently, probabilistic models have been proposed and used in vision. Szeliski's method has a few distinguishing features that make this monograph important and attractive. First, he presents a systematic Bayesian probabilistic estimation framework in which we can define and compute the prior model, the sensor model, and the posterior model. Second, his method represents and computes explicitly not only the best estimates but also the level of uncertainty of those estimates using second order statistics, i.e., the variance and covariance. Third, the algorithms developed are computationally tractable for dense fields, such as depth maps constructed from stereo or range finder data, rather than just sparse data sets. Finally, Szeliski demonstrates successful applications of the method to several real world problems, including the generation of fractal surfaces, motion estimation without correspondence using sparse range data, and incremental depth from motion.

The work reported here represents an important progress toward a systematic treatment of uncertainty in vision. Yet, there is much more to be done, both theoretically and experimentally. I believe this research monograph will be a rich source of references and ideas for further research in this important area.

Takeo Kanade
May 2, 1989

Preface

Over the last decade, many low-level vision algorithms have been devised for extracting depth from intensity images. The output of such algorithms usually contains no indication of the uncertainty associated with the scene reconstruction. The need for such error modeling is becoming increasingly recognized, however. This modeling is necssary because of the noise which is inherent in real sensors and the need to optimally integrate information from different sensors or viewpoints.

In this book, which is a revision of my 1988 Ph.D. thesis at Carnegie Mellon University, I present a Bayesian model which captures the uncertainty associated with low-level vision. My model is applicable to two-dimensional dense fields such as depth maps, and it consists of three components: a prior model, a sensor model, and a posterior model. The prior model captures any *a priori* information about the structure of the dense field. I construct this model by using the smoothness constraints from regularization to define a Markov Random Field. The sensor model describes the behavior and noise characteristics of the measurement system. A number of sensor models are developed for both sparse depth measurements and dense flow and intensity measurements. The posterior model combines the information from the prior and sensor models using Bayes' Rule. This model can then be used as the input to later stages of processing. I show how to compute optimal estimates from the posterior model and also how to compute the uncertainty (variance) in these estimates.

This book describes the application of Bayesian modeling to a number of low-level vision problems. The main application is the on-line extraction of depth from motion, for which I use a two-dimensional generalization of the Kalman filter. The resulting incremental algorithm provides a dense on-line estimate of depth which is initially crude, but whose uncertainty and error are reduced over time. In other applications of Bayesian modeling, I show how the Bayesian interpretation of regularization can be used to choose the optimal smoothing parameter for interpolation; I develop a Bayesian model which estimates observer motion from sparse depth measurements without correspondence; and I use the fractal nature of the prior model to construct multiresolution relative surface representations. The uncertainty modeling techniques that I develop, and the utility of these techniques in various applications, demonstrate the usefulness of Bayesian modeling for low-level vision.

The work in this book owes its genesis and success to the support of many colleagues and friends. In particular, I would like to thank my thesis advisors, Geoffrey Hinton and Takeo Kanade, for their ideas, encouragement and guidance. During my five year stay at Carnegie Mellon, they helped me discover the joys of research, the necessity for clear and positive presentation, and the excitement of interaction with peers and colleagues. I thank the other members of my thesis committee, Jon Webb and Alex Pentland, for their comments on the thesis and their additional ideas and insights.

I thank the members of the IUS vision research group at Carnegie Mellon, especially Steve Shafer, Martial Hebert, Larry Matthies, and In So Kweon, for many interesting discussions on computer vision problems. I am also grateful to the members of the Boltzmann research group, especially David Plaut, for providing an interesting alternative perspective on many perception problems.

I thank Steve Zucker for first firing my interest in vision, and for many subsequent discussions. I also thank the computer vision researchers at other institutions, who both during visits to CMU and during discussions at conferences have provided many interesting ideas, insights and suggestions.

I am grateful for the outstanding graduate research environment which exists in the Computer Science Department at Carnegie Mellon. I thank all of the faculty, administrators and students whose efforts contribute to this environment. I also thank of all the many friends in the department who made my stay in Pittsburgh so enjoyable.

Since my graduation from Carnegie Mellon, I have been fortunate to work with two outstanding research groups, first at Schlumberger Palo Alto Research and then at SRI International. At Schlumberger, I had interesting discussions with Jay Tenenbaum, and was fortunate to collaborate with Demetri Terzopoulos on a number of ideas, and to learn from him the intricacies of physically-based modeling and numerical relaxation. At SRI International, I am grateful for the support and guidance of Martin Fischler, and for stimulating interactions with Yvan Leclerc and the other members of the perception group.

Most of all, I thank my wife, Lyn Lovell McCoy, whose love and support over the last few years have brought joy to my daily endeavors and made the completion of this work possible.

This research was sponsored in part by the Defense Advanced Research Projects Agency (DOD), ARPA Order No. 5976 under Contract F33615-87-C-1499 and monitored by: Avionics Laboratory, Air Force Wright Aeronautical Laboratories, Aeronautical Systems Division (AFSC), Wright-Patterson AFB, OH 45433-6543. Support was also given by an Allied Corporation scholarship and by the National Science Foundation under Grant Number IRI-8520359, and by Schlumberger and SRI International.

Richard Szeliski
May 11, 1989

BAYESIAN MODELING OF UNCERTAINTY
IN LOW-LEVEL VISION

Chapter 1

Introduction

This book examines the application of Bayesian modeling to low-level vision. Bayesian modeling is a probabilistic estimation framework that consists of three separate models. The prior model describes the world or its properties which we are trying to estimate. The sensor model describes how any one instance of this world is related to the observations (such as images) which we make. The posterior model, which is obtained by combining the prior and sensor models using Bayes' Rule, describes our current estimate of the world given the data which we have observed.

The main thesis of this book is that Bayesian modeling of low-level vision is both feasible and useful. In the course of the book, we will develop a number of Bayesian models for low-level vision problems such as surface interpolation and depth-from-motion. To show that our approach is feasible, we will build computationally tractable Bayesian models using Markov Random Fields. To show that these models are useful, we will develop representations and algorithms that yield significant improvements in terms of capabilities and accuracy over existing regularization and energy-based low-level vision algorithms.

The computationally tractable versions of Bayesian models that we use involve estimating first and second order statistics. The first order statistics of a probability distribution are simply its *mean* values. Many low-level vision algorithms already perform this estimation, either by explicitly using Bayesian models, or by using optimization to find the best estimate (the best and mean estimates often coincide in these vision problems). The second order statistics, which encode the *uncertainty* or the *variance* in these estimates, are used much less frequently. The application of uncertainty estimation in computer vision and robotics has thus far been limited to systems that have a small number of parameters, such as the position and orientation of a mobile robot, or the location of discrete features. In this book, we will extend uncertainty modeling to dense correlated fields, such as the depth or optical flow maps that are commonly used in low-level vision.

In the introduction, we begin by examining what characterizes low-level vision problems and why the modeling of uncertainty in this context is important. We follow this discussion with a brief survey of previous research in low-level vision and other related areas which forms the background for our work. The main results contained in this monograph are then summarized, followed by the organization of the remainder of this book.

1.1   Modeling uncertainty in low-level vision

Low-level visual processing is often characterized as the extraction of *intrinsic images* (Barrow and Tenenbaum 1978) such as depth, orientation or reflectance from the input intensity images (Figure 1.1). A characteristic of these images is that they often represent *dense fields*, i.e., the information is available at all points in the two-dimensional visual field. This dense, retinotopic information is then segmented and grouped into coherent surfaces, parts and objects by later stages of processing.

Intrinsic images form a useful intermediate representation and facilitate the task of higher level processing. Intrinsic characteristics such as depth or reflectance are more useful than raw intensities for scene understanding or object recognition since they are closer to the true physical characteristics of the scene. Intrinsic images provide a more stable description than intensities, one that does not change, say, with illumination. These intermediate representations also provide a framework for integrating information from multiple low-level vision modules such as stereo, shading, occluding contours, motion, and texture, and for integrating information over time.

Much of the processing that occurs in these early stages of vision deals with the solution of *inverse problems* (Horn 1977). The physics of image formation confounds many different phenomena such as lighting, surface reflectivity, surface geometry, and projective geometry. Low-level visual processing attempts to recover some or all of these features from the sampled image array by making assumptions about the world being viewed. For example, when solving the surface interpolation problem—the determination of a dense depth map from a sparse set of depth values—the assumption is made that surfaces vary smoothly in depth (except at object or part boundaries). In Chapter 4 we will argue that many of these assumptions can be viewed as *intrinsic models*—probabilistic assumptions about the form of likely intrinsic images.

The inverse problems arising in low-level vision are generally *ill-posed* (Poggio and Torre 1984), since the data insufficiently constrains the desired solution. One approach to overcoming this problem, called *regularization* (Tikhonov and Arsenin 1977), imposes additional weak smoothness constraints in the form of stabilizers. Another approach, *Bayesian modeling* (Geman and Geman 1984,

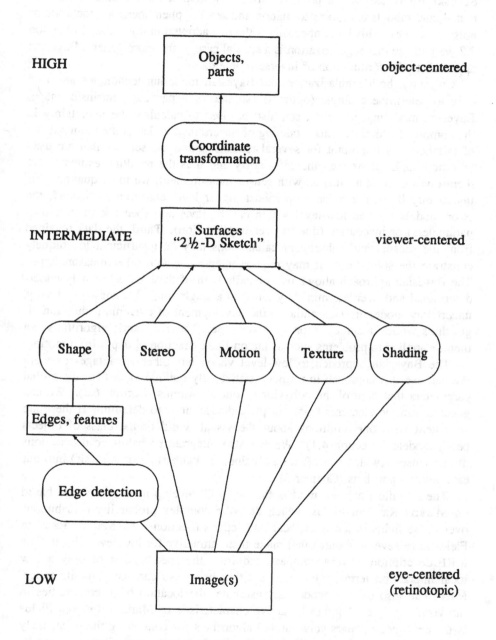

Figure 1.1: Visual processing hierarchy

Szeliski 1986), assumes a prior statistical distribution for the data being estimated, and models the image formation and sensing phenomena as stochastic or noisy processes. This latter approach is the one adopted in this book. In Section 3.2 we will see that regularization is a special case of the more general Bayesian approach to the formulation of inverse problems.

Currently, both regularization and Bayesian modeling techniques are used only to determine a single (optimal) estimate of a particular intrinsic image. Bayesian modeling, however, can also be used to calculate the uncertainty in the computed estimate. This modeling of uncertainty, which is the main subject of this book, is important for several reasons. First, the sensors that are used in vision applications are inherently noisy and thus the resulting estimates are themselves uncertain. Just as with sensor measurement, we must quantify this uncertainty if we are to develop robust higher level algorithms. Second, the prior models used in low-level vision can be uncertain (because of unknown parameters) or inaccurate (due to oversimplification). Third, the data obtained from the sensors and subsequent calculations may be insufficient to uniquely constrain the solution, or it may require integration with other measurements. The Bayesian approach allows us to handle both of these cases, namely underdetermined and overdetermined systems, in a single unified framework. Lastly, uncertainty modeling is essential to the development of dynamic estimation algorithms whose accuracy improves over time. These dynamic algorithms can then be applied to problems such as the on-line extraction of depth from motion.

The Bayesian approach to low-level vision has other advantages as well. We can use this approach to estimate statistically optimal values for the global parameters that control the behavior of our algorithms (Section 6.3). We can generate sample elements from our prior distributions to determine if these are consistent with our intuitions about the visual world or the class of objects being modeled (Section 4.1). We can also integrate probabilistic descriptions of our sensors (which can often be obtained by calibration or analysis) into our estimation algorithms (Chapter 5).

The specific Bayesian models that we will develop in this book are based on Markov Random Fields, which describe complex probability distributions over dense fields in terms of local interactions (Section 3.2). Markov Random Fields have several features that make them attractive for low-level vision. The MRF description is very compact, requiring the specification of only a few local interaction terms, while the resulting behaviors can have infinite range. MRFs can also easily encode (and estimate) the location of discontinuities in the visual surface. Another attractive characteristic of Markov Random Fields is the existence of massively parallel algorithms for computing the most likely or mean estimates. As we will see in Section 6.2, these same algorithms can be used for computing uncertainty estimates. Using these algorithms, we can thus directly exploit the increasing parallelism that is becoming available in image processing architectures.

Unfortunately, iterative parallel algorithms sometimes converge very slowly towards the desired estimate. Multiresolution techniques can be used to speed up the convergence. These techniques, which operate on a pyramidal representation of the field, have previously been applied to deterministic relaxation algorithms such as those used in conjunction with regularization. In this book, we will extend these techniques to stochastic algorithms, such as those that are used in uncertainty estimation. We will also develop algorithms that use a *relative representation*. In such a representation, each level in a multiresolution hierarchy encodes the local difference in depth at an appropriate scale, and the sum of all the levels represents the absolute depth map. The advantage of this decomposition is that we can develop multiresolution relaxation algorithms that operate on all of the levels in parallel, and that we produce a multiscale description of the surface.

To summarize, intrinsic images are a useful intermediate representation for disambiguating different physical characteristics that are confounded during image formation and for providing a more stable description for higher levels of processing. By modeling the uncertainty associated with these intrinsic maps, we can cope with noisy sensors, develop a uniform integration framework, and process dynamic data. Using a Bayesian model also allows us to estimate global parameters, to study our prior models, and to incorporate probabilistic sensor models. The particular Bayesian models that we will use in this book are Markov Random Fields combined with multiresolution representations and algorithms. Before discussing these ideas in more detail, let us briefly review some of the relevant previous work.

1.2   Previous work

The research described in this book has been inspired by several different threads of research running through the fields of computer vision, artificial intelligence and estimation theory. In this section, we briefly mention some of this relevant research, tracing the development of ideas in the field. First, we review a number of general representations, computational theories and algorithms used in vision and related fields. Second, we examine some specific low-level vision problems and describe how the theories can be applied to their solution. The ideas presented in this section are discussed in more detail later in the book, particularly in Chapters 2 and 3.

The formulation of low-level vision as a transformation from input images to intermediate representations was first advocated by Barrow and Tenenbaum (1978) and Marr (1978).   Marr (1982) particularly stressed the importance of representations in developing theories about vision. Marr's idea of a "2½-

D sketch" was further formalized when Terzopoulos (1984) proposed visible surface representations as a uniform framework for interpolating and integrating sparse depth and orientation information. More recently, Blake and Zisserman (1987) have suggested that discontinuities in the visible surface are the most stable and important features in the intermediate level description, a view that seems to be echoed by Poggio *et al.* (1988). These and other representational issues are discussed in more detail in Chapter 2.

The computational theories used in conjunction with these surface representations were formulated first in terms of variational principles by Grimson (1983) and Terzopoulos (1983), then later formalized using regularization theory (Poggio *et al.* 1985a, Terzopoulos 1986a). Several methods have been proposed for discontinuity detection, including continuation (Terzopoulos 1984), Markov Random Fields (Marroquin 1984), weak continuity constraints (Blake and Zisserman 1986b), and minimum length encoding (Leclerc 1989). Similar energy-based models have also been extended to full three-dimensional modeling by Terzopoulos *et al.* (1987).

The common element in these computational theories is the minimization of a global energy function composed of many local energy components. This minimization has usually been implemented using iterative algorithms. The earliest cooperative algorithms were applied to the stereo matching problem (Julesz 1971, Dev 1974, Marr and Poggio 1976). A different class of iterative algorithms called relaxation labeling (Waltz 1975, Rosenfeld *et al.* 1976, Hinton 1977) was used to find solutions to symbolic constraint satisfaction problems. The idea of constraint propagation for numerical problems was first suggested by Ikeuchi and Horn (1981), and has been used in many subsequent low-level vision algorithms (Horn and Schunck 1981, Hildreth 1982, Grimson 1983). Multigrid methods (Terzopoulos 1983), which are based on multiple resolution representations (Rosenfeld 1980), have been used to speed up the convergence of numerical relaxation.

A common problem with relaxation algorithms is that they can find only locally optimal solutions, instead of finding the global minimum of the energy function. To overcome this, one possible strategy is to use coarse-to-fine matching algorithms (Marr and Poggio 1979, Witkin *et al.* 1987), which can be based on a multiresolution description of the signal (Witkin 1983). Another solution is to use a stochastic minimization technique called simulated annealing (Kirkpatrick *et al.* 1983, Hinton and Sejnowski 1983). This latter approach can also be related to Bayesian modeling through the use of Markov Random Fields (Geman and Geman 1984, Marroquin 1985). Deterministic approximations to simulated annealing can also be used (Koch *et al.* 1986), and these algorithms can be implemented using neural networks (Hopfield 1982, Ackley *et al.* 1985, Rumelhart *et al.* 1986). Continuation methods have also been investigated (Terzopoulos 1988, Blake and Zisserman 1987, Leclerc 1989). An alternative to these cooperative algorithms is to use global search (Ohta and Kanade 1985).

The application of Bayesian modeling to low-level vision has received relatively little attention. Markov Random Field models have been used to characterize piecewise constant images (Geman and Geman 1984) or surfaces (Marroquin 1985). Error models have been developed for stereo matching (Matthies and Shafer 1987) and for more abstract sensors (Durrant-Whyte 1987). The use of different loss functions to derive alternative optimal posterior estimators has been studied by Marroquin (1985). In the domain of real-time processing of dynamic data, the Kalman filter has been applied to the tracking of sparse features such as points or lines (Faugeras *et al.* 1986, Rives *et al.* 1986), and has recently been extended to dense fields (Matthies *et al.* 1987). In this book, we will extend the Bayesian approach to low-level vision by analyzing the uncertainty inherent in dense estimates, and by developing a number of new algorithms based on the Bayesian framework.

The representations and algorithms just described have been applied to a variety of low-level vision problems. Three that are central to this book are stereo, motion, and surface interpolation. Additional low-level vision problems include shape from contour (Barrow and Tenenbaum 1981, Kanade 1981), shape from shading (Ikeuchi and Horn 1981, Woodham 1981), and shape from texture (Witkin 1981). While these latter applications are not examined in this book, the same uncertainty modeling techniques that are developed here can be applied to many of these problems.

Stereo matching is one of the earliest vision problems to have been studied using a computational approach (Barnard and Fischler 1982). Early algorithms focused on the correspondence problem (Marr 1982), using random-dot stereograms as their test domain (Julesz 1971, Dev 1974, Marr and Poggio 1976). Subsequent algorithms have used zero-crossings (Marr and Poggio 1979, Grimson 1981, Kass 1984), edges (Henderson *et al.* 1979, Baker 1982, Arnold 1983, Ohta and Kanade 1985) or image intensities (Lucas 1984, Witkin *et al.* 1987) as the basic matching primitives. A variety of computational characterizations of the task have been used, including local support (Prazdny 1985, Szeliski and Hinton 1985, Drumheller and Poggio 1986), edge similarity (Arnold 1983, Kass 1984), segment similarity (Ohta and Kanade 1985) and regularization (Poggio *et al.* 1985a, Witkin *et al.* 1987). Among the solution methods used are cooperative algorithms (Marr and Poggio 1976), coarse-to-fine matching (Grimson 1981, Witkin *et al.* 1987), dynamic programming (Ohta and Kanade 1985), iterative improvement (Lucas 1984), and simulated annealing (Szeliski 1986, Barnard 1986). Despite the great diversity among stereo algorithms, some common characteristics do exist. Stereo correspondence is usually cast as an energy minimization problem, where the energy encodes both the feature compatibilities (similarity) and the smoothness or coherence in the estimated depth map. Since the energy function is usually non-convex, search techniques such as relaxation, coarse-to-fine matching, global search or stochastic optimization must be used. For stereo matching, the imaging geometry is usually known, thus reducing this search to a one-dimensional problem along epipolar lines in the two images (Ohta and Kanade 1985). While the range of target matches is thus restricted, the search itself usually cannot be performed independently on each line because of the two-dimensional smoothness constraint on the disparity field.

Extracting depth from motion resembles stereo matching in many ways, since similar features, matching criteria and algorithms can be used. Often, however, the epipolar lines are not known, and two dimensional flow vectors must be estimated. Optical flow can be computed by calculating the ratio of spatial and temporal gradients (Horn and Schunck 1981) or using correlation (Anandan 1984). Since the motion between frames is usually small, the optical flow estimates can be quite noisy, and must be smoothed using area-based (Horn and Schunck 1981, Anandan and Weiss 1985) or contour-based (Hildreth 1982) methods. Smoothing is also necessary to overcome the aperture problem (Hildreth 1982), although correlation-based flow estimators can avoid this ambiguity near corners and in textured areas (Anandan 1984). While optical flow has usually been measured from successive pairs of images, more recent motion algorithms have attempted to use the whole image sequence, by either fitting lines to the spatio-temporal data (Bolles and Baker 1985), using spatio-temporal filtering (Adelson and Bergen 1985, Heeger 1986), or using Kalman filtering (Matthies *et al.* 1987). In the context of this book, motion processing has three interesting characteristics. First, since the motions involved are often small, the correspondence problem is reduced. Second, the reliability of the flow information varies spatially, and hence this information must be smoothed. Third, the time varying nature of the data provides a good application for dynamic Bayesian models such as the Kalman filter.

Surface interpolation is often seen as a post-processing stage that integrates the sparse output of independent low-level vision modules (Marr 1982), although it has recently been used in conjunction with other algorithms such as stereo (Hoff and Ahuja 1986, Chen and Boult 1988). Surface interpolation was first studied in the context of stereo vision (Grimson 1981). An interpolation algorithm based on variational principles was developed by Grimson (1983), then extended to use multiresolution computation by Terzopoulos (1983), and finally reformulated using regularization (Poggio *et al.* 1985a). Recent research has focused on detecting discontinuities in the visible surface, using Markov Random Fields (Marroquin 1984), continuation methods (Terzopoulos 1986b), and weak continuity constraints (Blake and Zisserman 1986a). Determining the optimal amount of smoothing has also been investigated by Boult (1986) (see also Section 6.3). In the early parts of this book, surface interpolation will be studied as an isolated system. In Chapter 7, we will show how it can be integrated with other low-level vision modules.

As we have seen in this section, visible surface representations, energy-based models, cooperative computation, multiresolution algorithms, and Bayesian modeling are all themes that will play an important role in our presentation. In this book, we will develop a Bayesian model of visible surfaces based on regularization and Markov Random Fields. This model will allow us to compute the uncertainty in the surface estimate and to integrate new data over time. The algorithms that we develop will use massively parallel cooperative computation and multiresolution techniques. We will apply our new model to surface interpolation, and later to motion and stereo. With this background in mind, let us turn to the specific new results in this book.

1.3 Overview of results

This book contains both theoretical and experimental results. Its main contribution to computer vision theory is the development of a new probabilistic framework for modeling dense fields and their associated uncertainties. This framework is used to study the prior models defined by regularization, the sensor models used by low-level vision algorithms, and the posterior estimates obtained as the output of such vision algorithms. These models extend the theory of low-level visual processing and are used to develop a number of new low-level vision algorithms. The experimental results that we present use both synthetic and real scenes and are used to quantify the performance of these new algorithms. Additional details about these contributions are presented below.

The most important result in this book is the development of a Bayesian model for the dense fields that are commonly used in low-level vision. This approach is based on the observation that the estimates obtained with regularization and other energy-based minimization algorithms are equivalent to the posterior estimates obtained with Markov Random Field modeling. While this viewpoint is becoming more commonly accepted (Poggio *et al.* 1988), probabilistic modeling of low-level vision has so far only been used as a better tool for obtaining a single deterministic estimate (Marroquin 1985)[1]. The Bayesian framework has *not* previously been used to study the properties of the prior models, to better characterize sensor characteristics, or to treat the posterior estimate as a probabilistic quantity.

In developing our Bayesian framework, we start by interpreting regularization as Bayesian estimation, and by interpreting the stabilizer that enforces smoothness in the solution as defining probabilistic prior model. While this prior model *favors* smooth surfaces, it does not say how smooth *typical* sur-

---

[1]In the related area of connectionist modeling, however, Hinton and Sejnowski (1983) have used estimated co-occurrence statistics to develop a learning algorithm for the Boltzmann Machine.

faces will be. Using Fourier analysis, we re-write the energy of the stabilizer in terms of the power spectrum of the surface. Since the energy of the surface is related to its probability through the Gibbs distribution (Section 3.2), we show that the prior model is correlated Gaussian noise with a power spectrum defined by the choice of stabilizer. For the membrane and the thin plate, the two most commonly used regularization models, this spectrum is fractal (Szeliski 1987). This result leads us to a new understanding of the role of regularization: the choice of a particular stabilizer (degree of smoothness) is equivalent to assuming a particular power spectrum for the prior model.

Our development of the regularization-based probabilistic prior model allows us to construct two new algorithms. The first is a computer graphics algorithm that uses multigrid stochastic relaxation to generate fractal surfaces (Szeliski and Terzopoulos 1989a). These surfaces can be arbitrarily constrained—with depth and orientation constraints and depth and orientation discontinuities—and thus exhibit a degree of flexibility not present in previous algorithms. Our second result is the construction of a relative multiresolution representation. We show how the power spectrum of the composite representation can be computed by summing the power spectra of the individual levels. This gives us a new technique for shaping the frequency response characteristics of each level, and for ensuring the desired global smoothing behavior.

We also apply probabilistic modeling to the sensors used in low-level vision. We start by reviewing the equivalence between a point sensor with Gaussian noise and a simple spring constraint, and show how to extend this model to other one-dimensional uncertainty distributions. We then develop a new sensor model that incorporates the full three-dimensional uncertainty in a sparse depth measurement. The constraint corresponding to this model acts like a force field, and is related to the elastic net model of Durbin and Willshaw (1987). We also show how Bayesian modeling can be used to analyze the uncertainty in a correlation-based optical flow estimator and to develop an error model for a simple imaging system.

After developing the prior and sensor models, we examine the characteristics of the posterior model. We show how the uncertainty in the posterior estimate can be calculated from the energy function of the system, and we devise two new algorithms to perform this computation. The first algorithm uses deterministic relaxation to calculate the uncertainty at each point separately. The second algorithm generates typical random samples from the posterior distribution, and calculates statistics based on these samples (Figure 1.2).

The probabilistic description of the posterior estimate allows us to construct two new parameter estimation algorithms. The first algorithm estimates the optimal amount of smoothing to be used in regularization. This is achieved by maximizing the likelihood of the data points that were observed given a particular (parameterized) prior model. A similar Bayesian model is used to

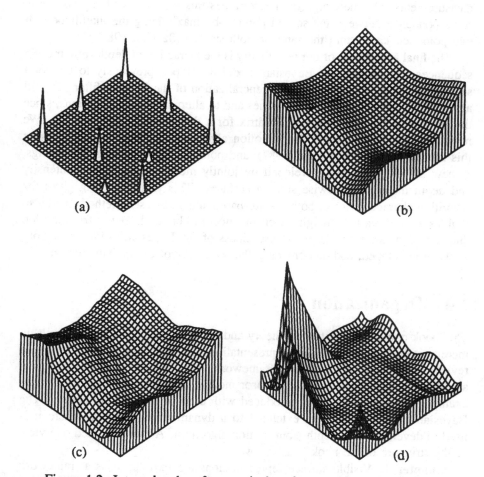

Figure 1.2: Interpolated surface, typical surface, and uncertainty map
This figure shows pictorially two of the most important ideas in this book. The nine data points in (a) are smoothed using regularization to obtain surface (b). While (b) shows the estimated *mean* value of the surface from which the points were taken, (c) shows a *typical* surface from which they might have come (this surface is locally fractal). By generating a number of such surfaces, we can estimate the local uncertainty in the interpolated solution at each point on the surface (d). This uncertainty, which tells us about the local variance in the estimate, has been magnified for easier interpretation. Note how the uncertainty is higher near the corners where the plate has more wobble and is also high near the tear in the surface.

determine observer or object motion given two or more sets of sparse depth measurements. Our new algorithm determines this motion—without using any correspondence between the sensed points—by maximizing the likelihood that two point sets came from the same smooth surface (Szeliski 1988a).

The final application of our new theory is the extraction of depth from motion sequences. We extend the Bayesian model developed previously to temporal sequences using a two-dimensional generalization of the Kalman filter. Careful attention is paid to computational issues and to alternative representations, since the storage of the full covariance matrix for a dense field is not feasible. We review an incremental depth-from-motion algorithm that was developed using this framework (Matthies *et al.* 1987), and point out some of its shortcomings. A new algorithm is then developed by jointly modeling the current intensity and depth maps as piecewise smooth surfaces. This joint modeling gives the algorithm characteristics of both feature-based and area-based methods, and also enables a more realistic imaging (sensor) model to be used. Depth-from-motion thus serves as an example of the usefulness of the Bayesian estimation theory we have developed, and demonstrates the feasibility of its implementation.

1.4   Organization

This book contains a mixture of theory and applications. The theory is built up incrementally, starting with the representations used for low-level vision and a review of the Bayesian modeling framework. This framework is then instantiated by developing prior models, sensor models, and posterior models. Simple examples of applications are introduced with each of these three models. The Bayesian framework is then extended to a dynamic environment, where it is used to develop a new depth-from-motion algorithm. A more detailed overview of the structure of the book is as follows:

Chapter 2. Visible surface representations are introduced as a framework for sensor integration and dynamic vision. A discrete implementation of this representation—based on finite element analysis—is presented, and the cooperative solution of regularized problems is explained. Multiresolution representations are then reviewed, followed by an introduction to relative representations and a discussion of hierarchical basis functions. Finally, the role and implementation of discontinuities is discussed, along with alternative visual representations.

Chapter 3. The division of Bayesian models into prior models, sensor models, and posterior models is explained in the context of low-level vision. Markov Random Fields are introduced, and some basic mathematical results are reviewed. The utility of probabilistic and MRF models for low-level vision is discussed.

Chapter 4. The special role of prior models in low-level vision is presented. The stabilizers used in regularization are related to prior models through the Gibbs distribution. Fourier analysis of the resulting prior is used to show that these models are fractal, and to suggest a spectral domain interpretation for the choice of stabilizer. The application of these results to the generation of constrained fractals is described. The application of the spectral analysis to the construction of relative representations is presented. The chapter closes with a discussion of mechanical vs. probabilistic models.

Chapter 5. This chapter examines both sparse measurements, such as those obtained from stereo matching or range finders, and dense measurements, as available from optical flow or raw intensity images. First, the relationship of Gaussian noise to spring models is reviewed. A new sensor model based on three dimensional Gaussian noise is then introduced, and its properties are examined. An error model for correlation-based optical flow measurements is developed. Finally, a simple probabilistic model of an image sensor (CCD camera) is presented.

Chapter 6. By combining a prior model and an appropriate sensor model we obtain a posterior model of the visible surface. Maximum A Posteriori (MAP) estimation is introduced along with alternative optimal estimates. Two new methods for calculating the uncertainty in these estimates are then described. The probabilistic framework is used to derive a new method for estimating the regularization (smoothing) parameter. This same framework is also used to develop a new algorithm for motion estimation that does not require correspondence between sensed points.

Chapter 7. The framework developed in Chapter 3 is extended to work with temporally varying data using the Kalman filter. An incremental depth-from-motion algorithm based on this framework (Matthies *et al.* 1987) is reviewed. Motivated by certain shortcomings in this algorithm, a new algorithm is proposed that explicitly models the estimated intensity and disparity fields and localizes discontinuities to sub-pixel accuracy.

Chapter 8. The final chapter summarizes the main results in this work and discusses open questions and areas of future research.

Appendices. These contain derivations of mathematical results used in the body of the book: (A) a description of the finite element discretization and multigrid relaxation algorithm; (B) the Fourier analysis of both the prior model and the stochastic posterior estimation algorithms; (C) an error model for correlation-based optical flow estimation; (D) the equations for the posterior estimate and the marginal distribution of the data.

Chapter 2

Representations for low-level vision

Representations play a central role in the study of any visual processing system (Marr 1982). The *representations* and *algorithms* that describe a visual process are a particular instantiation of a general *computational theory*, and are constrained by the *hardware* that is available for their implementation. Representations make certain types of information explicit, while requiring that other information be computed when needed. For example, a depth map and an orientation map may represent the same visible surface, but complex computations may be required to convert from one representation to the other. The choice of representation becomes crucial when the information being represented is uncertain (McDermott 1980).

In this chapter, we will examine representations suitable for modeling visible surfaces. In the context of the hierarchy of visual processing (Figure 1.1), these representations are actually at the interface between the low and intermediate stages of vision. We will first review retinotopic visible surface representations, and discuss the difference between continuous and discrete fields. We will then examine the use of regularization, finite element analysis and relaxation for specifying and solving low-level vision problems. This is followed by an introduction to multigrid algorithms and relative multiresolution representations. The modeling and localization of discontinuities in the surface is then examined. Lastly, we mention a number of alternative representations used in low-level computer vision.

2.1   Visible surface representations

The visible surface representations that we develop in this book are related to Marr's *2½-dimensional (2½-D) sketch* (Marr 1978) and Barrow and Tenenbaum's *intrinsic images* (Barrow and Tenenbaum 1978). The 2½-D sketch is a retinotopic map that encodes local surface orientation and distance to the viewer, as well as discontinuities in the orientation and distance maps. Intrinsic images represent scene characteristics such as distance, orientation, reflectance and illumination in multiple retinotopic maps. The role of these intermediate representations in visual processing is indicated in Figure 2.1.

Visible surface representations are dense retinotopic maps that can be modeled as functions over a two-dimensional domain aligned with the image. Theoretically, these fields are continuous, with the domain being a subset of $\mathbf{R}^2$. In practice, discrete fields are used, with the domain being a subset of $\mathbf{Z}^2$. One possible method for deriving these discrete approximations from the continuous representation is finite element analysis, which we examine in the next section.

Visible surface representations can be used to integrate the output of different vision modules or different sensors (Figure 2.1). They can also be used to integrate information from different viewpoints and to fill in or smooth out information obtained from low-level processes. One possible technique for performing this integration and interpolation is regularization, which we examine in the next section. Another technique is Markov Random Field modeling, which we examine in Section 3.2. Both of these approaches are examples of intrinsic models (Chapter 4)—locally parameterized models of shape that describe surfaces at an intermediate level before their segmentation or grouping into parts or objects.

Before proceeding with a description of these techniques, we should briefly discuss the question: "Are visible surface representations necessary?" The early work on intermediate representations (Barrow and Tenenbaum 1978, Marr 1978) was motivated by a disappointment with feature-based approaches to vision and a desire to incorporate computational models of image formation. Some of the recent research in computer vision, however, has suggested that image features can be grouped and matched directly to a model (Lowe 1985) or to a more general parts description (Pentland 1986).

Psychophysical studies and recent computational modeling suggest that both models of visual processing (hierarchical and direct) are present in human vision and can be used in computer vision applications. The ability to obtain depth perception from random-dot stereograms (Julesz 1971) strongly suggests an independent stereo vision module that produces an intermediate depth map. Studies in neurophysiology show the existence of multiple visual maps in the cortex (Van Essen and Maunsell 1983). These multiple maps may be the structure used by intermediate level processes involving visual attention and pre-attentive grouping. In this book, we will concentrate on the formation and representation of intermediate level maps and ignore for now the problems associated with higher levels of visual processing.

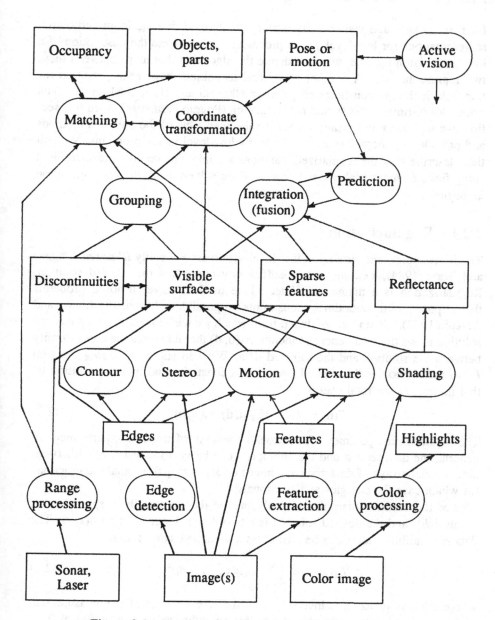

Figure 2.1: A more complex visual processing hierarchy

2.2   Visible surface algorithms

Intensity images and visible surface representations define the input and output representations for low-level vision processes. To complete the description of a low-level vision module, we must define the algorithm that maps between these two representations. A number of general techniques have been proposed for this task, including constraint propagation (Ikeuchi and Horn 1981), variational principles (Grimson 1983), and regularization (Poggio *et al.* 1985a). In this section, we will use regularization since it subsumes most of the previous methods and provides a general framework for many low-level vision problems. We will then describe how the regularized equations are implemented on a discrete field using finite element analysis and how they are solved using iterative relaxation algorithms.

2.2.1   Regularization

The inverse problems arising in low-level vision are generally *ill-posed* (Poggio and Torre 1984), i.e., the data insufficiently constrains the desired solution. Regularization is a mathematical technique used to solve ill-posed problems that imposes weak smoothness constraints on possible solutions (Tikhonov and Arsenin 1977). Given a set of data **d** from which we wish to recover a regularized solution **u**, we define an energy function $E_d(\mathbf{u}, \mathbf{d})$ that measures the compatibility between the solution and the sampled data. We then add a *stabilizing* function $E_p(\mathbf{u})$ that embodies the desired smoothness constraint and find the solution $\mathbf{u}^*$ that minimizes the total energy

$$E(\mathbf{u}) = (1 - \lambda)E_d(\mathbf{u}, \mathbf{d}) + \lambda E_p(\mathbf{u}). \tag{2.1}$$

The regularization parameter λ controls the amount of smoothing performed. In general, the data term **d** and the solution **u** can be vectors, discrete fields (two-dimensional arrays of data such as images or depth maps), or analytic functions (in which case the energies are functionals).

For the surface interpolation problem, the data is usually a sparse set of points $\{d_i\}$, and the desired solution is a two-dimensional function $u(x, y)$. The data compatibility term can be written as a weighted sum of squares

$$E_d(\mathbf{u}, \mathbf{d}) = \frac{1}{2} \sum_i c_i[u(x_i, y_i) - d_i]^2, \tag{2.2}$$

where the confidence c_i is inversely related to the variance of the measurement d_i, i.e., $c_i = \sigma_i^{-2}$. Two examples of possible smoothness functionals (taken from Terzopoulos (1984)) are the *membrane* model

$$E_p(\mathbf{u}) = \frac{1}{2} \int \int \left(u_x^2 + u_y^2 \right) dx \, dy, \tag{2.3}$$

which is a small deflection approximation of the surface area, and the *thin plate* model

$$E_p(\mathbf{u}) = \frac{1}{2} \int \int \left(u_{xx}^2 + 2u_{xy}^2 + u_{yy}^2 \right) \, dx \, dy, \tag{2.4}$$

which is a small deflection approximation of the surface curvature (note that here the subscripts indicate partial derivatives). These two models can be combined into a single functional

$$E_p(\mathbf{u}) = \frac{1}{2} \int \int \rho(x,y)\{[1 - \tau(x,y)][u_x^2 + u_y^2] + \tau(x,y)[u_{xx}^2 + 2u_{xy}^2 + u_{yy}^2]\} \, dx \, dy \tag{2.5}$$

where $\rho(x,y)$ is a *rigidity* function, and $\tau(x,y)$ is a *tension* function . The rigidity and tension functions can be used to allow depth ($\rho(x,y) = 0$) and orientation ($\tau(x,y) = 0$) discontinuities. The minimum energy solutions of systems that use the above smoothness constraint are "generalized piecewise continuous splines under tension" (Terzopoulos 1986b).

As an example, consider the nine data points shown in Figure 2.2a. The regularized solution using a thin plate model is shown in Figure 2.2b. We can also introduce a depth discontinuity along the left edge and an orientation discontinuity along the right to obtain the solution shown in Figure 2.2c.

The stabilizer $E_p(\mathbf{u})$ described by (2.5) is an example of the more general controlled-continuity constraint

$$E_p(\mathbf{u}) = \frac{1}{2} \sum_{m=0}^{p} \int w_m(\mathbf{x}) \sum_{j_1 + \cdots + j_d = m} \frac{m!}{j_1! \cdots j_d!} \left| \frac{\partial^m u(\mathbf{x})}{\partial x_1^{j_1} \cdots \partial x_d^{j_d}} \right|^2 \, d\mathbf{x} \tag{2.6}$$

where $\mathbf{x}$ is the (multidimensional) domain of the function u. A generalized version of the data compatibility term is

$$E_d(\mathbf{u}, \mathbf{d}) = \frac{1}{2} \int c(\mathbf{x}) |u(\mathbf{x}) - d(\mathbf{x})|^2 d\mathbf{x} \tag{2.7}$$

where $d(\mathbf{x})$ and $c(\mathbf{x})$ are now continuous functions. These general formulations will be used in Appendix B in order to analyze the filtering behavior of the regularized solution.

Regularization has been applied to a wide variety of low-level vision problems (Poggio and Torre 1984). In addition to surface interpolation, it has been used for shape from shading (Horn and Brooks 1986), stereo matching (Barnard 1986, Witkin *et al.* 1987), and optical flow (Anandan and Weiss 1985). Problems such as surface interpolation and optical flow smoothing have a quadratic energy function, and hence have only one local energy minimum. Other problems, such as stereo matching, may have many local minima and require different algorithms for finding the optimum solution (Szeliski 1986, Barnard 1986, Witkin *et al.* 1987).

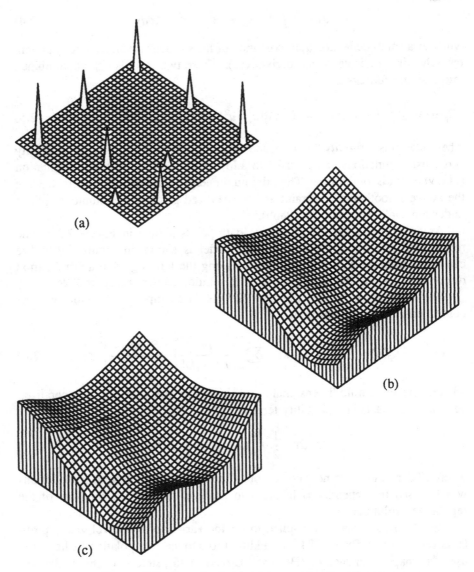

Figure 2.2: Sample data and interpolated surface
(a) data points (b) interpolated thin plate solution (c) thin plate solution with a
tear along the left edge and a crease along the right edge

2.2.2 Finite element discretization

To find the minimum energy solution on a digital or analog computer, it is necessary to discretize the domain of the solution $u(\mathbf{x})$ using a finite number of *nodal variables*. The usual and most flexible approach is to use finite element analysis (Terzopoulos 1984). In this work, we will restrict our attention to rectangular domains on which a rectangular fine grained mesh has been applied. The topology of this mesh is fixed and does not depend on the location of the data points. It can thus be used for integrating data from various sensors or from various viewpoints. The fine grained nature of the mesh leads to a natural implementation on a massively parallel array of processors. This kind of massively parallel network is similar to the visual processing architecture of the retina and primary visual cortex.

As an example, let us examine the finite element approximation for the surface interpolation problem. Using a triangular conforming element for the membrane and a non-conforming rectangular element for the thin plate (Terzopoulos 1984), we can derive the energy equations

$$E_p(\mathbf{u}) = \frac{1}{2} \sum_{(i,j)} [(u_{i+1,j} - u_{i,j})^2 + (u_{i,j+1} - u_{i,j})^2] \tag{2.8}$$

for the membrane (the subscripts indicate spatial position) and

$$E_p(\mathbf{u}) = \frac{1}{2} h^{-2} \sum_{(i,j)} [\ (u_{i+1,j} - 2u_{i,j} + u_{i-1,j})^2 \\ + 2(u_{i+1,j+1} - u_{i,j+1} - u_{i+1,j} + u_{i,j})^2 \\ + (u_{i,j+1} - 2u_{i,j} + u_{i,j-1})^2] \tag{2.9}$$

for the thin plate, where $h = |\Delta x| = |\Delta y|$ is the size of the mesh (isotropic in x and y). These equations hold at the interior of the surface. Near border points or discontinuities some of the energy terms are dropped or replaced by lower continuity terms (see Appendix A). The equation for the data compatibility energy is simply

$$E_d(\mathbf{u}, \mathbf{d}) = \frac{1}{2} \sum_{(i,j)} c_{i,j}(u_{i,j} - d_{i,j})^2 \tag{2.10}$$

with $c_{i,j} = 0$ at points where there is no input data.

If we concatenate all the nodal variables $\{u_{i,j}\}$ into one vector $\mathbf{u}$, we can write the prior energy model as one quadratic form

$$E_p(\mathbf{u}) = \frac{1}{2} \mathbf{u}^T A_p \mathbf{u}. \tag{2.11}$$

This quadratic form is valid for any controlled-continuity stabilizer, though the coefficients will differ. The stiffness[1] matrix A_p is typically very sparse, but it

---

[1] This term comes from the finite element analysis of structures.

is not tightly banded because of the two-dimensional structure of the field. The rows of A_p are fields of the same dimensionality and extent as the discretized field x and can be described in terms of computational molecules (Terzopoulos 1988). For the membrane and thin plate, typical molecules are

$$
\begin{bmatrix} & -1 & \\ -1 & 4 & -1 \\ & -1 & \end{bmatrix} \quad \text{and} \quad h^{-2} \begin{bmatrix} & & 1 & & \\ & 2 & -8 & 2 & \\ 1 & -8 & 20 & -8 & 1 \\ & 2 & -8 & 2 & \\ & & 1 & & \end{bmatrix}.
$$

The A_p matrix is analogous to the weight matrix of a connectionist network.

For the data compatibility model we can write

$$
E_d(\mathbf{u}, \mathbf{d}) = \frac{1}{2}(\mathbf{u} - \mathbf{d})^T A_d (\mathbf{u} - \mathbf{d}) \tag{2.12}
$$

where A_d is usually diagonal (for uncorrelated sensor noise) and may contain zeros along the diagonal. The resulting overall energy function $E(\mathbf{u})$ is quadratic in $\mathbf{u}$

$$
E(\mathbf{u}) = \frac{1}{2}\mathbf{u}^T A \mathbf{u} - \mathbf{u}^T \mathbf{b} + c \tag{2.13}
$$

with

$$
A = A_p + A_d \quad \text{and} \quad \mathbf{b} = A_d \mathbf{d}. \tag{2.14}
$$

The energy function has a minimum at

$$
\mathbf{u}^* = A^{-1}\mathbf{b} \tag{2.15}
$$

and can thus be re-written as

$$
E(\mathbf{u}) = \frac{1}{2}(\mathbf{u} - \mathbf{u}^*)^T A (\mathbf{u} - \mathbf{u}^*) + k. \tag{2.16}
$$

2.2.3 Relaxation

Once the parameters of the energy function have been determined, we can calculate the minimum energy solution $\mathbf{u}^*$ using relaxation. This approach has two advantages over direct methods such as Gaussian elimination or triangular decomposition. First, direct methods do not preserve the sparseness of the A matrix, and thus require more than just a small amount of storage per node. Second, relaxation methods can be implemented on massively parallel locally connected computer architectures (or even on analog networks (Koch *et al.*

1986)). A number of relaxation methods such as Jacobi, Gauss-Seidel, successive overrelaxation (SOR), and conjugate gradient have been used for visible surface interpolation[2].

For our simulations on a sequential machine, we use Gauss-Seidel relaxation where nodes are updated one at a time. This method is simple to implement, converges faster than the parallel Jacobi method, and can easily be converted to a stochastic version known as the Gibbs Sampler (see Section 4.2). At each step, a selected node is set to the value that locally minimizes the energy function. The energy function for node u_i (with all other nodes fixed) is

$$E(u_i) = \frac{1}{2}a_{ii}u_i^2 + (\sum_{j \in N_i} a_{ij}u_j - b_i)u_i + k, \qquad (2.17)$$

where the subscripts i and j are actually two-element vectors that index the image position. The node value that minimizes this energy is therefore

$$u_i^+ = \frac{b_i - \sum_{j \in N_i} a_{ij}u_j}{a_{ii}}. \qquad (2.18)$$

Note that it is possible to use a parallel version of Gauss-Seidel relaxation so long as nodes that are dependent (have a non-zero a_{ij} entry) are not updated simultaneously. This parallel version can be implemented on a mesh of processors for greater computational speed.

The result of applying this iterative algorithm on the nine data points of Figure 2.2a are shown in Figure 2.3 for 10, 100 and 1000 iterations. As can be seen, this relaxation algorithm converges very slowly towards the optimal solution. This slow convergence may not be a problem in a dynamic system (Chapter 7) where iteration can proceed in parallel with data acquisition and the system can converge to an adequate solution over the course of time[3]. However, for one-shot interpolation problems, the convergence speed may be more critical, and the multigrid techniques described in the next section may be required.

2.3 Multiresolution representations

Multiresolution pyramids are representations that can be used for efficiently solving many image processing tasks (Rosenfeld 1984). Pyramids have also been used for generating multiscale descriptions of signals such as intensity images

---

[2]Terzopoulos (1984) presents a brief survey of relaxation methods. Gauss-Seidel is used by Terzopoulos (1984), while conjugate gradient descent is used by Choi (1987), and SOR by Blake and Zisserman (1987).

[3]Once the system has latched-on to a good solution, it can then track changes in the scene using only a few iterations to correct its estimate. Similar arguments have recently been advanced in support of *dynamic deformable models* by Terzopoulos (1987) and Kass *et al.* (1988).

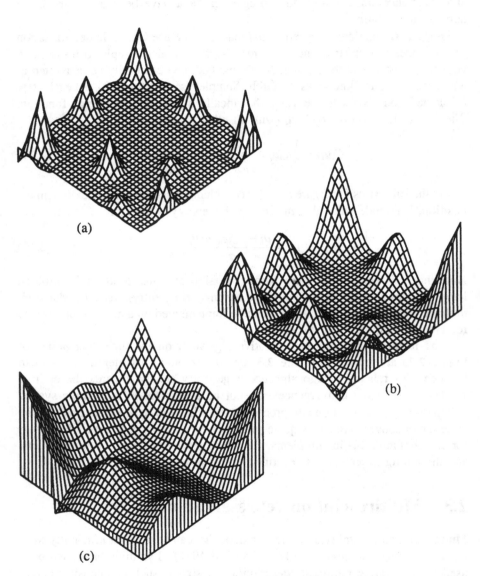

Figure 2.3: Single level relaxation algorithm solutions

(a) after 10 iteration (b) 100 iterations (c) 1000 iterations

(Burt and Adelson 1983, Crowley and Stern 1982, Mallat 1987). The use of multiresolution representations and multigrid algorithms for visible surface interpolation was first extensively studied by Terzopoulos (1984). In this section, we will review these multigrid algorithms and then introduce relative representations as an alternative to regular multiscale descriptions. The full development of these relative representations will have to wait until Chapter 4 where we present a probabilistic interpretation of surface representations. We will also present hierarchical basis functions as an alternative to multigrid relaxation.

2.3.1 Multigrid algorithms

Multigrid relaxation algorithms (Hackbusch and Trottenberg 1982, Hackbusch 1985) are based on the observation that local iterative methods are good at reducing the high frequency components of the interpolation error, but are poor at reducing the low frequency components. By solving the same problem on a coarser grid, this low frequency error can be reduced more quickly. Multigrid algorithms thus operate on a hierarchy of resolution levels as shown in Figure 2.4.

To develop a multigrid algorithm, several components must be specified:

- the number of levels and the size of the grid at each level

- the method used to derive the energy equations at the coarser levels from the fine level equations

- a *restriction* operation that maps a solution at a fine level to a coarser grid

- a *prolongation* operation that maps from the coarse to the fine level

- a *coordination scheme* that specifies the number of iterations at each level and the sequence of prolongations and restrictions.

A sophisticated version of multigrid relaxation was developed by Terzopoulos (1984). In his implementation, the coarser meshes are coincident with the original mesh, i.e., the coarser level nodes are a subset of the finer level nodes. Simple injection (subsampling) is used for restriction, and third-degree Lagrange interpolation is used for prolongation. A Full Multigrid (FM) method is used to coordinate the inter-level interactions. This ensures that the coarse levels have the same accuracy as the fine level solutions.

In this work, we will use a simpler version, since optimal relaxation algorithms and convergence rates are not our primary concern. Averaging is used for restriction and bilinear interpolation for prolongation. The coordination scheme is a simple coarse-to-fine algorithm, where the prolongated coarse solution is used as a starting point for the next finer level. Figure 2.5 shows the regularized

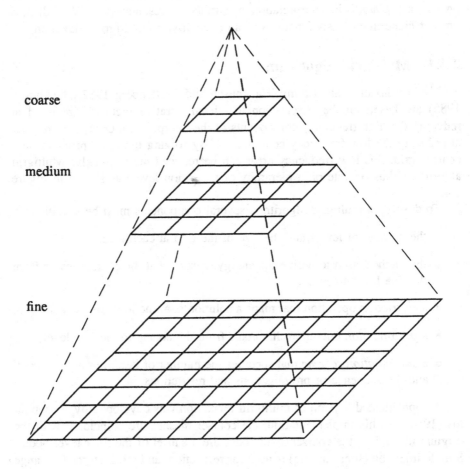

coarse

medium

fine

Figure 2.4: Multiresolution pyramid

solution at the coarse, medium and fine resolution levels for the example that was introduced previously.

To derive the energy functions at the coarser level, the discontinuities and data points are first mapped between levels using simple block averaging. The new energy equations are then derived in the same fashion as for the finest level. Appendix A gives the details of the multigrid algorithm which we have adopted, and Appendix B contains a Fourier analysis of its convergence properties.

2.3.2 Relative representations

The multigrid techniques presented in the previous section keep a full copy of the current depth estimate at each level[4]. Relaxation is applied to each level separately, and projection operators are used to map between levels. This computational framework thus fails to exploit the full parallelism inherent in a pyramidal representation. Some attempts have been made to implement fully parallel relaxation algorithms using this multigrid representation (Terzopoulos 1985), but they have met with limited success.

An alternative to the absolute multiresolution representation just described is a *relative* representation. In this representation, each level encodes details pertinent to its own scale, and the *sum* of all the levels provides the current depth estimate. The relative representation is thus analogous to a band-pass image pyramid (Burt and Adelson 1983, Crowley and Stern 1982, Mallat 1987), while the absolute representation is similar to a low-pass pyramid. In contrast to these band-pass representations, the relative representation which we develop here is *not* obtained by repeated filtering of a fine-resolution map. Instead, each level in the pyramid has its own associated prior energy (weak smoothness constraint). The data compatibility is measured in terms of the distance between the summed depth map and the data points. The interpolated solution is then obtained by finding the minimum energy configuration of the whole system. We will present a more detailed description of this approach later in this section.

The relative multiresolution representation offers several possible advantages over the usual absolute representation. Fully parallel relaxation can be used with this representation, and yields a multiscale decomposition of the visible surface. Discontinuities can be assigned to just one level, thus permitting a better description of the scene. For example, in a *blocks world* scene (an indoors robotics environment with several objects lying on a table), the fine level may describe the shape of an individual object, while the coarse level may describe the table elevation. The use of relative representations can also increase the descriptive power of a method when uncertainty is being modeled (McDermott 1980). For example, knowing that a particular block sitting on a table is 10 ± 0.9

---

[4]In the multigrid literature, this is called the *Full Approximation* (FA) scheme.

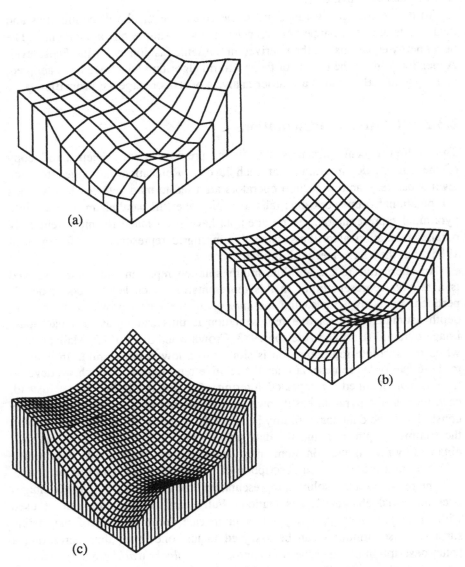

Figure 2.5: Coarse-to-fine multiresolution relaxation solution

(a) coarse level (b) medium level (c) fine level

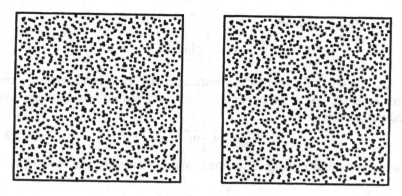

Figure 2.6: Random-dot stereogram showing Cornsweet illusion

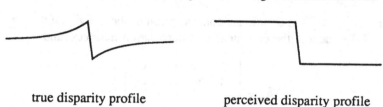

true disparity profile   perceived disparity profile

Figure 2.7: Disparity profiles for random-dot stereogram

cm tall is more precise than knowing that the block top is 80 ± 4.1 cm above the floor and that the table top is 70 ± 4 cm above the floor[5].

A relative representation also permits us to incorporate relative depth measurements. An example of such a measurement is the disparity gradients available from stereopsis when the vergence angle is not known. A similar effect is apparent in the emergence of depth perception from random-dot stereograms. Figure 2.6, which was suggested by Geoffrey Hinton, shows a stereo pair whose disparity profile is the same as that used in the Cornsweet illusion (see also Anstis and Howard (1978)). Using a stereo viewer or by cross-eyed viewing of the stereogram, the reader may elicit a perception of a surface in depth. The first and strongest perception is that of a step edge, which is due to the large disparity gradient in the center of the image (Figure 2.7). This illusion also suggests that multiresolution channels may be responsible for disparity processing. These channels would feed naturally into a relative multiresolution representation. The exact details of this mechanism remain to be worked out.

To explore the possibility of using a relative representation, we will write down the equations for this approach and try a simple example. Using $\mathbf{u}_l$ to represent the depth map at level l, we can write the overall depth map as a sum

---

[5]For this example, we add variances instead of standard deviations to obtain the uncertainties.

of the individual level maps

$$\mathbf{u} = \sum_{l=1}^{L} \mathbf{I}_l \mathbf{u}_l$$

where $\mathbf{I}_l$ is the interpolation function associated with each level (note that the $\mathbf{u}_l$ get smaller as we go up the pyramid). Each level has its own associated prior (smoothness) energy

$$E_p^l(\mathbf{u}_l) = \frac{1}{2} \mathbf{u}_l^T A_p^l \mathbf{u}_l. \tag{2.19}$$

The data compatibility energy is defined using the summed representation $\mathbf{u}$,

$$E_d(\mathbf{u}, \mathbf{d}) = \frac{1}{2}(\mathbf{u} - \mathbf{d})^T A_d(\mathbf{u} - \mathbf{d}). \tag{2.20}$$

Using $\tilde{\mathbf{u}} = [\mathbf{u}_1^T \ldots \mathbf{u}_L^T]^T$ to denote the concatenation of the individual state vectors, $\tilde{\mathbf{I}} = [\mathbf{I}_1 \ldots \mathbf{I}_L]$ to denote the concatenated interpolation matrices, and

$$\tilde{A}_p = \begin{bmatrix} A_p^1 & 0 & \cdots & 0 \\ 0 & A_p^2 & \cdots & 0 \\ \vdots & \vdots & \ddots & \vdots \\ 0 & 0 & \cdots & A_p^L \end{bmatrix}$$

to denote the composite prior energy matrix, we can write the overall energy function as

$$\begin{aligned} E(\tilde{\mathbf{u}}) &= E_d(\tilde{\mathbf{u}}, \mathbf{d}) + \sum_{l=1}^{L} E_p^l(\mathbf{u}_l) \\ &= \frac{1}{2}(\tilde{\mathbf{I}}\tilde{\mathbf{u}} - \mathbf{d})^T A_d(\tilde{\mathbf{I}}\tilde{\mathbf{u}} - \mathbf{d}) + \frac{1}{2}\tilde{\mathbf{u}}^T \tilde{A}_p \tilde{\mathbf{u}} \\ &= \frac{1}{2}\tilde{\mathbf{u}}^T \tilde{A}\tilde{\mathbf{u}} - \tilde{\mathbf{u}}^T \tilde{\mathbf{b}} + c \end{aligned} \tag{2.21}$$

where

$$\tilde{A} = \tilde{\mathbf{I}}^T A_d \tilde{\mathbf{I}} + \tilde{A}_p \quad \text{and} \quad \tilde{\mathbf{b}} = \tilde{\mathbf{I}}^T A_d \mathbf{d}. \tag{2.22}$$

The minimum energy solution can be calculated from this quadratic form as

$$\tilde{\mathbf{u}}^* = \tilde{A}^{-1}\tilde{\mathbf{b}} = (\tilde{\mathbf{I}}^T A_d \tilde{\mathbf{I}} + \tilde{A}_p)^{-1}\tilde{\mathbf{I}}^T A_d \mathbf{d}. \tag{2.23}$$

Designing a set of spline energies for each level that decompose $\mathbf{u}$ into a reasonable multiresolution description and also have a desired global smoothing behavior is nontrivial. If we try something simple, such as setting A_p^l at each level to the same value as would be used in the regular multigrid algorithm (e.g., those values specified in Appendix A), we find that the matrix $\tilde{A}$ is singular which precludes a unique relative representation solution. This occurs because $E_p^l(\mathbf{u}_l)$

is invariant with respect to additive constants, so that adding a constant value to one level and subtracting it from another affects neither the solution energy nor the summed solution (assuming that the interpolators reproduce constant functions).

To overcome this problem, we can add a small quadratic energy term to each level, i.e., increment $\mathbf{A}_p^l$ by $\epsilon\mathbf{I}$, where $\mathbf{I}$ is the unit matrix. Since we also want the larger variations in depth to be encoded by the coarser level, we reduce the magnitude of the energy as we go up the pyramid by setting

$$\mathbf{A}_p^l = s^{-l}(\mathbf{A}_p^l + \epsilon), \quad \text{with } s > 1,$$

where $\mathbf{A}_p^l$ is derived using the usual finite element analysis (Appendix A). Figure 2.8 shows the three level relative decomposition and the summed solution using this approach (with $s = 4$).

The summed depth map shown in Figure 2.8d obviously differs from the absolute solution shown in Figure 2.2b. One reason for this is that bilinear interpolation used in Figure 2.8 is not sufficiently smooth to be used with a thin plate. More generally, the effective smoothing behavior of the relative representation is different than that of the absolute representation.

To compute the exact smoothing behavior, we examine the minimum energy solution written in terms of the prior energy matrices $\mathbf{A}_p^l$, the data energy matrix $\mathbf{A}_d$, and the data points $\mathbf{d}$. From (2.23), we see that the overall summed depth solution is

$$\mathbf{u}^* = [\bar{\mathbf{I}}(\bar{\mathbf{I}}^T\mathbf{A}_d\bar{\mathbf{I}} + \tilde{\mathbf{A}}_p)^{-1}\bar{\mathbf{I}}^T]\mathbf{A}_d\mathbf{d}. \tag{2.24}$$

For this to be equivalent to $\mathbf{u}^* = (\mathbf{A}_p + \mathbf{A}_d)^{-1}\mathbf{A}_d\mathbf{d}$, the minimum of the original energy equation (2.13), we must have

$$\bar{\mathbf{I}}(\bar{\mathbf{I}}\mathbf{A}_d\bar{\mathbf{I}} + \tilde{\mathbf{A}}_p)^{-1}\bar{\mathbf{I}}^T = (\mathbf{A}_p + \mathbf{A}_d)^{-1}.$$

Using the matrix inversion lemma (Appendix D) we can re-write this as

$$\bar{\mathbf{I}}(\tilde{\mathbf{A}}_p^{-1} - \tilde{\mathbf{A}}_p^{-1}\bar{\mathbf{I}}^T(\bar{\mathbf{I}}\tilde{\mathbf{A}}_p^{-1}\bar{\mathbf{I}}^T + \mathbf{A}_d^{-1})^{-1}\bar{\mathbf{I}}\tilde{\mathbf{A}}_p^{-1})\bar{\mathbf{I}}^T = \mathbf{A}_p^{-1} - \mathbf{A}_p^{-1}(\mathbf{A}_p^{-1} + \mathbf{A}_d^{-1})^{-1}\mathbf{A}_p^{-1}$$

which is satisfied if

$$\mathbf{A}_p^{-1} = \bar{\mathbf{I}}\tilde{\mathbf{A}}_p^{-1}\bar{\mathbf{I}}^T = \sum_{l=1}^{L}\mathbf{I}_l(\mathbf{A}_p^l)^{-1}\mathbf{I}_l^T. \tag{2.25}$$

We thus have a method for computing the global smoothing behavior of the relative representation in terms of the smoothing energy at each level and the interpolation functions[6]. If we can satisfy this condition, we can obtain a multiresolution description of our depth map while performing parallel relaxation, without changing the global smoothing properties of our interpolator.

---

[6]Equation (2.25) can also be derived by assuming that each level is a correlated Gaussian with covariance $(\mathbf{A}_p^l)^{-1}$ and calculating the covariance of the summed surface (see Section 4.3).

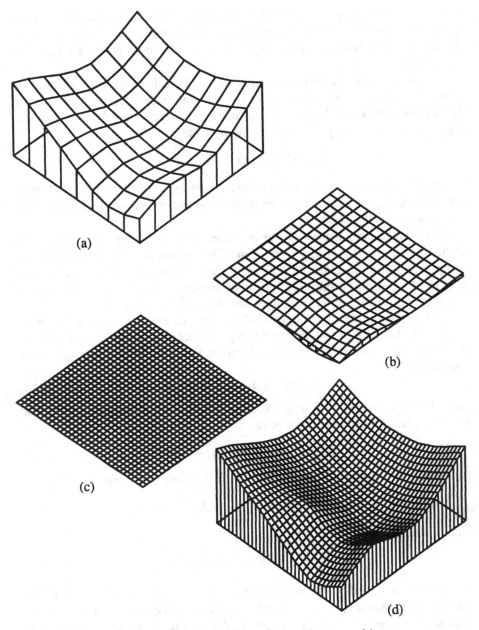

Figure 2.8: Relative multiresolution decomposition

(a) coarse level (b) medium level (c) fine level (d) summed solution

In practice, choosing a set of A_p^l that satisfy (2.25) for a given A_p (say a thin plate) and perform an adequate multiscale decomposition is difficult because while the A_p^l and A_p matrices are very sparse, their inverses are not (or they may not even exist). As we will see in Section 4.3, however, the smoothness constraints used in regularization are equivalent to assuming a particular spectral distribution for the prior model. This will allow us to re-cast the problem of satisfying (2.25) in terms of shaping the spectrum of each individual level such that the summed spectrum has the desired shape. We will therefore defer the full development of the relative representation until Section 4.3, where we can examine its construction using some new tools.

2.3.3 Hierarchical basis functions

The relative representation which we have developed is similar in structure to the hierarchical basis functions developed recently by Yserentant (1986). In this approach, the usual nodal basis set **u** is replaced by a hierarchical basis set **v**. Certain elements of the hierarchical basis set have larger support than the nodal basis elements, and this allows the relaxation algorithm to converge more quickly when using the hierarchical set.

To convert from the hierarchical to the nodal basis set, we use a simple linear (matrix) transform

$$\mathbf{u} = \mathbf{S}\mathbf{v} \quad \text{where} \quad \mathbf{S} = \mathbf{S}_1\mathbf{S}_2\ldots\mathbf{S}_{L-1} \tag{2.26}$$

and L is the number of levels in the hierarchical basis set. Each of the sparse matrices S_l interpolates the nodes at level $l+1$ to level l and adds in the nodes corresponding to the new level. The columns of S give the values of the hierarchical basis functions at the nodal variable locations.

In his paper, Yserentant uses recursive subdivision of triangles to obtain the nodal basis set. The corresponding hierarchical basis then consists of the top-level (coarse) triangularization, along with the subtriangles that are generated each time a larger triangle is subdivided. Linear interpolation is used on a triangle each time it is subdivided. We can generalize this notion to arbitrary interpolants defined over a rectangular grid (Szeliski 1989). Each node in the hierarchical basis is assigned to the level in the multiresolution pyramid where it first appears (Figure 2.9). This is similar to the relative representation, except that each level is only partially populated, and the total number of nodes is the same in both the nodal and hierarchical basis sets. To fully define the hierarchical basis set, we select an interpolation function that defines how each level is interpolated to the next finer level before the new node values are added in.

The resulting algorithms for mapping between the hierarchical and nodal basis sets are simple and efficient. We use

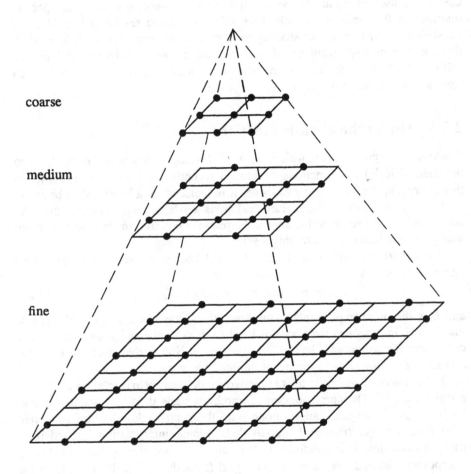

coarse

medium

fine

Figure 2.9: Hierachical basis pyramid

The circles indicate the nodes in the hierarchical basis.

```
procedure S
    for l = L − 1 down to 1
        for i ∈ M_l
            for j ∈ N_i
                u(i) = u(i) + w(i; j)u(j)
end S
```

to convert from the hierarchical to the nodal basis set. In this procedure, which goes from the coarsest level ($l = L$) to the finest ($l = 1$), each node is assigned to one of the level collections M_l. Each node also has a number of neighboring nodes N_i on the next coarser level that contribute to its value during the interpolation process. The $w(i; j)$ are the weighting functions that depend on the particular choice of interpolation function (these are the off-diagonal terms in the S_l matrices).

We will also use the adjoint of this operation

```
procedure S^T
    for l = 1 to L − 1
        for i ∈ M_l
            for j ∈ N_i
                u(j) = u(j) + w(i; j)u(i)
end S^T
```

in the conjugate gradient descent algorithm which we develop in the next section. An example of the hierarchical basis representation for the surface previously presented in Figure 2.2b is shown in Figure 2.10. In this representation, nodes that are coincident with higher level nodes have a 0 value.

These mapping algorithms are easy to code (once the interpolation functions have been precomputed) and require very few computations to perform. On a serial machine, these procedures use $O(n)$ operations (multiplications and additions), where n is the number of nodes. On a parallel machine, $O(L)$ parallel steps are required, where $L \approx \frac{1}{2} \log n$ is the number of levels in the pyramid. This is the same number of steps as is needed to perform the global summations (inner products) used in the conjugate gradient algorithm. Note that although we have defined the hierarchical basis over a pyramid, it can actually be represented in the same space as the usual nodal basis, and the transformations between bases can be accomplished in place.

The hierarchical basis set allows us to minimize exactly the same energy as the one we obtained from the discretization on the finest grid. Substituting $\mathbf{u} = \mathbf{Sv}$ into (2.13), we obtain the new energy equation

$$
\begin{aligned}
E(\mathbf{v}) &= \frac{1}{2}\mathbf{v}^T(\mathbf{S}^T\mathbf{AS})\mathbf{v} - \mathbf{v}^T(\mathbf{S}^T\mathbf{b}) + c \\
&= \frac{1}{2}\mathbf{v}^T\hat{\mathbf{A}}\mathbf{v} - \mathbf{v}^T\hat{\mathbf{b}} + c
\end{aligned}
\tag{2.27}
$$

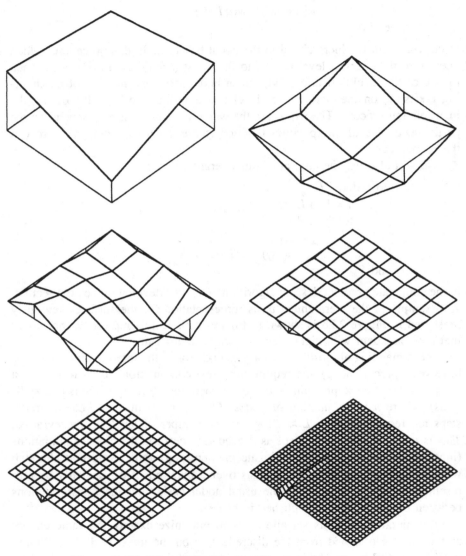

Figure 2.10: Hierarchical basis representation of solution

where the ˆ identifies the hierarchical basis vectors and matrices. The advantage of minimizing this new equation is that the condition number of the matrix $\hat{A}$ is much smaller than that of the original matrix A (Yserentant 1986), implying much faster convergence for iterative algorithms such as conjugate gradient.

Unfortunately, the $\hat{A}$ matrix is not as sparse as the original matrix A, so that a direct minimization of (2.27) is impractical. Instead, we use the recursive mappings S and S^T in conjunction with the original matrix A and vector b to compute the required residuals and inner products. The resulting conjugate gradient descent algorithm is identical to the usual single-level algorithm except that we use a smoothed version of the residual $\check{r} = SS^T r$ to choose the new direction (Szeliski 1989). We describe the development of this algorithm below.

Conjugate gradient descent is a numerical optimization technique closely related to steepest descent algorithms (Press *et al.* 1986). At each step k, a direction p_k is selected in the state space, and an optimal sized step is taken in this direction. In steepest descent, the direction is always equal to the current gradient of the function being minimized. In conjugate gradient descent, we modify this direction so that successive directions are *conjugate* with respect to A, i.e., $p_{k+1} A p_k = 0$. Figure 2.11 shows the result of applying conjugate gradient descent to the surface interpolation problem introduced in Figure 2.2.

A description of the usual (nodal basis) conjugate gradient descent is shown in the left column of Figure 2.12. Having selected a direction p_k, we choose the optimal step size α_k so as to minimize $\Delta E(u_k + \alpha_k p_k)$. This involves computing the product of the sparse matrix A and the p_k, and the inner product of the resulting vector w_k and p_k. On a fine-grained parallel architecture, the matrix operation is computable in constant time (dependent on the size of the neighborhoods or molecules in A) and the inner product summation is computable in $\log n$ steps using a summing pyramid. After updating the new state, we compute the new residual r_{k+1} and find the value of β_{k+1} which will make the new and old directions conjugate.

For a quadratic energy equation such as (2.13) or (2.27), the conjugate gradient algorithm is guaranteed to converge to the correct solution in n steps, in the absence of roundoff error. As we mentioned in the previous section, however, we can obtain much faster convergence to an approximate solution if we minimize (2.27) instead of (2.13). The resulting algorithm is shown in the right column of Figure 2.12. In this algorithm, we update the state of the hierarchical basis vector v_k by computing the residual vector $\hat{r}_k$ and direction vector $\hat{p}_k$. To implement the matrix multiplications $\hat{A}\hat{p}_k$ and $\hat{A}v_k$, we use the mapping operations S and S^T before and after the matrix product with the original sparse matrix A. In the process, we convert the quantities $\hat{p}_k$ and v_k into the nodal representations p_k and u_k. The overall algorithm thus uses two calls to S and two calls to S^T to compute the required quantities for the conjugate gradient descent.

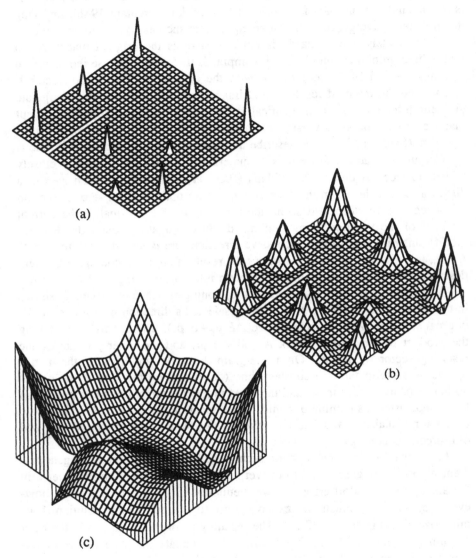

Figure 2.11: Conjugate gradient relaxation example

after (a) 1 iteration (b) 10 iterations (c) 100 iterations

Nodal basis conjugate gradient	Hierarchical basis conjugate gradient

0. initialize r_0 using 6.
   and set $p_0 = r_0$

1. $w_k = Ap_k$

2. $\alpha_k^D = p_k \cdot w_k = p_k^T Ap_k$
3. $\alpha_k^N = p_k \cdot r_k = p_k^T r_k$
4.† $\alpha_k = \alpha_k^N / \alpha_k^D$

5. $u_{k+1} = u_k + \alpha_k p_k$
6. $r_{k+1} = b - Au_{k+1}$

7. $\beta_{k+1}^N = r_{k+1} \cdot w_k = r_{k+1}^T Ap_k$
8. $\beta_{k+1} = \beta_{k+1}^N / \alpha_k^D$
9. $p_{k+1} = r_{k+1} - \beta_{k+1} p_k$

10. loop to 1.

0. initialize $\hat{r}_0$ using 6a. and 6b.
   and set $\hat{p}_0 = \hat{r}_0$

1a. $p_k = S\hat{p}_k$
1b. $w_k = Ap_k = AS\hat{p}_k$
1c. $\hat{w}_k = S^T w_k = (S^T AS)\hat{p}_k$

2. $\alpha_k^D = \hat{p}_k \cdot \hat{w}_k = p_k \cdot w_k$
3. $\alpha_k^N = \hat{p}_k \cdot \hat{r}_k = p_k \cdot r_k$
4. $\alpha_k = \alpha_k^N / \alpha_k^D$
5a. $v_{k+1} = v_k + \alpha_k \hat{p}_k$
5b. $u_{k+1} = Sv_{k+1} = u_k + \alpha_k p_k$
6a. $r_{k+1} = b - Au_{k+1}$
6b. $\hat{r}_{k+1} = S^T r_{k+1} = S^T b - (S^T AS)v_{k+1}$
6c. $\bar{r}_{k+1} = S\hat{r}_{k+1}$

7. $\beta_{k+1}^N = \hat{r}_{k+1} \cdot \hat{w}_k = \bar{r}_{k+1} \cdot w_k$
8. $\beta_{k+1} = \beta_{k+1}^N / \alpha_k^D$
9a. $\hat{p}_{k+1} = \hat{r}_{k+1} - \beta_{k+1} \hat{p}_k$
9b. $p_{k+1} = S\hat{p}_k = \bar{r}_{k+1} - \beta_{k+1} p_k$

10. loop to 1b.

† $\Delta E(u + \alpha p) = \frac{\alpha^2}{2} p^T Ap - \alpha p^T r$

Figure 2.12: Algorithms for nodal and hierarchical conjugate gradient descent

Inspection of the algorithm shown in Figure 2.12 shows that it can be simplified to reduce the number of mappings required. We note that in steps 2 and 3 the quantities α_k^N and α_k^D (the numerator and denominator of α_k) can be computed just as easily using the regular nodal basis representation. Upon reflection, this result is not surprising, since once the direction $\hat{\mathbf{p}}_k$ (or equivalently $\mathbf{p}_k$) has been selected, the size of the step must be the same independent of which representation is used. Similarly, in step 7 we can replace the inner product $\hat{\mathbf{r}}_{k+1} \cdot \hat{\mathbf{w}}_{k+1}$ with the inner product $\bar{\mathbf{r}}_{k+1} \cdot \mathbf{w}_{k+1}$, where $\bar{\mathbf{r}}_{k+1} = SS^T\mathbf{r}_{k+1}$ is the smoothed residual vector. If we now use step 5*b* instead of 5*a* and 9*b* instead of 9*a* and 1*a*, we obtain an algorithm that is nearly identical to the original conjugate gradient algorithm. The only difference is that we now smooth the residual vector $\mathbf{r}_{k+1}$ using a sweep up (S^T) and then back down (S) the pyramid to obtain the vector $\bar{\mathbf{r}}_{k+1}$. This smoothed vector dictates the new direction.

To evaluate the performance of our new algorithm, we ran a number of experiments on synthetic data sets. The set of points used in each run is the one shown in Figure 2.2a. For each of the three models tested (membrane, thin plate, and controlled-continuity thin plate), the optimal solution $\mathbf{u}^*$ for the 33×33 grid was computed by using 2000 iterations of conjugate gradient descent. For each experiment (defined by a suitable choice of interpolator and number of levels), the root mean squared (RMS) error

$$e_k = |\mathbf{u}_k - \mathbf{u}^*|/\sqrt{n}$$

was plotted as a function of the number of iterations k.

Figure 2.13 shows the state of the hierarchical conjugate gradient algorithm ($L = 4$) after 1, 10, and 100 iterations. A dramatic speedup over both Gauss-Seidel relaxation (Figure 2.3) and conjugate gradient (Figure 2.11) is evident. Figure 2.14 shows the effects of varying the number of smoothing levels in the pyramid (L) on the convergence rate. The topmost curve ($L = 1$) is the convergence rate of the usual nodal basis conjugate gradient descent algorithm. The surprising result is that the fastest convergence is obtained when $L = 4$ and $L = 5$, instead of $L = 6$ (the full pyramid). Initially, the larger number of smoothing levels used leads to a faster convergence. As time goes on, however, the large number of levels tends to over-smooth the residual vector. The optimum number of levels to be used seems to be related to the density of the underlying data points, but this remains to be verified empirically. Another possibility, which remains to be investigated, is to adjust the number of levels adaptively during the relaxation.

Figure 2.15 shows the effects of using different interpolators on the convergence rate (for this plot, the best value of L, usually 4 or 5, was used). The bilinear interpolator and bilinear interpolator with discontinuities seem to work the best. The convergence rates for the thin plate without discontinuities

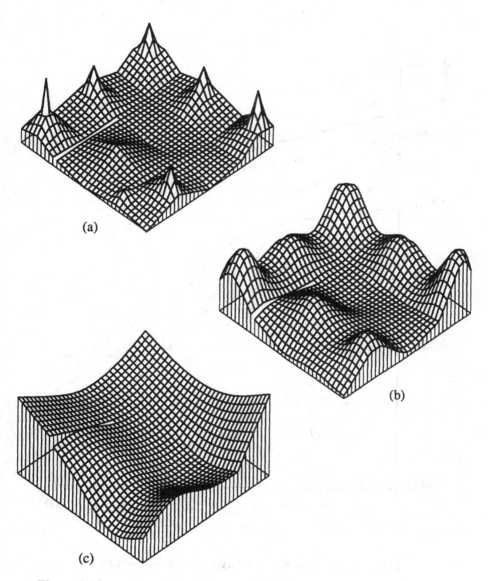

Figure 2.13: Hierarchical conjugate gradient ($L = 4$) relaxation example

after (a) 1 iteration (b) 10 iterations (c) 100 iterations

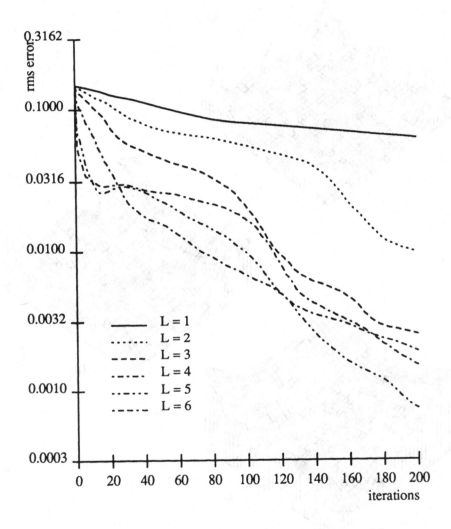

Figure 2.14: Algorithm convergence as a function of L

Controlled-continuity thin plate, bilinear interpolator.

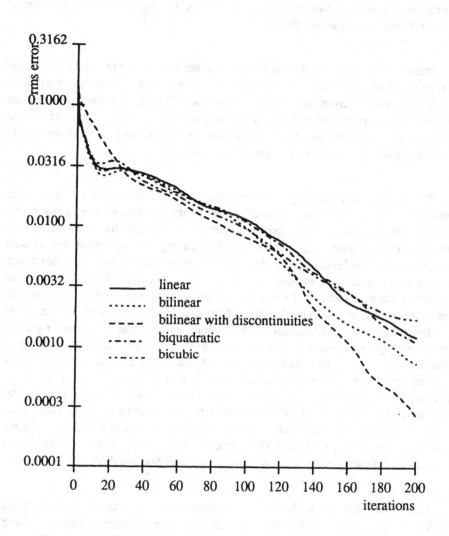

Figure 2.15: Algorithm convergence as a function of interpolator

Controlled-continuity thin plate, $L = 4$ or 5.

and continuous membrane are even faster. Compared to coarse-to-fine Gauss-Seidel relaxation, the hierarchical conjugate gradient algorithm is much faster. A comparison with full-multigrid (Hackbusch 1985) is currently being performed (Szeliski and Terzopoulos 1989b).

From the experiments described here, we see that the hierarchical basis conjugate gradient algorithm is dramatically faster than single-resolution conjugate gradient (which is itself much faster than Gauss-Seidel). Similar speedups are available using multigrid techniques (Hackbusch 1985). The biggest advantage of this new approach over multigrid techniques is the ease of implementation and its suitability for massively parallel architectures.

When designing a multigrid algorithm, we must first devise a hierarchy of problems, i.e., for each level, we must re-derive the finite element equations (this often involves averaging data from the finer level). We also have to specify both the injection (subsampling) and prolongation (interpolation) operations, as well as choose an inter-level coordination scheme. With hierarchical basis functions, only a single interpolation function needs to be specified. There is no need to explicitly build a pyramid for representation or computation (this is also true for some multigrid techniques). Most of the computation proceeds in parallel at the fine level, with only occasional excursions up or down the virtual pyramid for summing or smoothing. Since we can choose the interpolation function (hierarchical basis) independent of the problem being solved, we have much greater flexibility. For example, wavelets (Mallat 1987) could be used as an alternative to the polynomial bases which we have studied in this paper. The hierarchical basis function idea can easily be extended to domains other than two-dimensional surfaces. It could just as easily be applied to 3-D elastic models that use cylindrical coordinates (Terzopoulos *et al.* 1987), or to 3-D elastic net models built from a recursively tessellated sphere (Section 4.4).

Comparing the hierarchical basis approach to the relative representation which we introduced previously, we see that main difference between these two representations is the amount of state utilized. The relative representation uses more state than the original fine-level description, whereas the hierarchical basis uses the same amount. To constrain the extra degrees of freedom in the relative representation, we must use a separate smoothness functional for each level. This makes it more difficult to match the original smoothing characteristics of the fine-level equations, but also gives us more control over the exact frequency characteristics. With the hierarchical basis representation, the exact same energy equation as was used with the fine level is minimized.

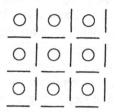

Figure 2.16: Dual lattice for representing discontinuities
Depth values are represented by circles and discontinuities by line segments.

2.4 Discontinuities

Representing and localizing discontinuities is an important component of the study of surface interpolation and other low-level vision processes. The detection of intensity discontinuities (*edge detection*) has a long history, dating back to the earliest days of the computer vision field (Roberts 1965, Hueckel 1971); it remains an active area of research (Marr and Hildreth 1980, Canny 1986, Leclerc and Zucker 1987). The estimation of depth and orientation discontinuities in parallel with surface interpolation was first studied by Terzopoulos (1984).

In the finite element representation developed by Terzopoulos, depth and orientation discontinuities are located on the same grid as the depth values themselves. More recent implementations (Geman and Geman 1984, Marroquin 1984, Harris 1987, Blake and Zisserman 1987) use a *dual lattice* for depth discontinuities (Figure 2.16). We will adopt this latter representation in this work, with the addition of orientation discontinuities coincident with the original depth nodes (see Appendix A). The quantization of discontinuity curves represented on the dual lattice is far from ideal. As observed by Blake and Zisserman (1987), diagonal curves have a longer apparent length in this representation, and are thus penalized more in the detection phase. More sophisticated representations of curves are currently being developed (Zucker 1986), but have yet to be integrated with visible surface representations.

Several different methods have been developed for detecting discontinuities in parallel with the surface interpolation process. Continuation methods, which gradually introduce discontinuities at locations of high curvature, have been investigated by Terzopoulos (1984). Markov Random Fields have been used in conjunction with stochastic optimization by Geman and Geman (1984) and Marroquin (1984). A deterministic approximation to these stochastic algorithms that uses analog "neural nets" was studied by Koch *et al.* (1986). Weak continuity constraints, which are similar to Markov Random Field descriptions, have been used with the graduated non-convexity (GNC) algorithm by Blake and Zisserman (1986b). The use of intensity edges for constraining the location of depth discontinuities has been studied by Gamble and Poggio (1987).

The accurate localization of depth discontinuities is an important element of visible surface estimation. Without discontinuities, regularization-based methods tend to over-smooth the data, and the accuracy of the reconstruction is reduced. Discontinuity detection can also be combined with surface segmentation (Leclerc 1989), which is an important first step in higher level analysis. It has even been recently suggested that discontinuities in the visible surface are more important than the depth values themselves (Blake and Zisserman 1987, Poggio *et al.* 1988).

When intensity edges are used for motion tracking or stereo matching, the precision of the edge position affects the accuracy of the estimated depth of the feature (Matthies and Shafer 1987). Edge detectors that localize the edge position to sub-pixel precision (Nalwa 1986) are extremely useful in this context. Sub-pixel localization of discontinuities in a depth map can also be useful, especially in a multiresolution hierarchy. In such a representation, the sub-pixel position of a discontinuity at a coarse level can be used to modify the prolongation operation (coarse-to-fine interpolation), thus providing a better initial fine level solution. The coarse level discontinuity position can be determined unambiguously from the fine level so long as the discontinuities are sufficiently far apart.

The research presented in this book largely ignores the problem of discontinuity detection and localization because of their inherent complexity. The treatment of discontinuities is limited in the earlier parts of the book to the introduction of discontinuities by hand to illustrate the flexibility of prior models and the effects of discontinuities on uncertainty maps. In Chapter 7, we show how the sub-pixel localization of intensity edges and depth discontinuities can be used in an incremental depth-from-motion algorithm to significantly improve the accuracy of the depth estimates. The automatic detection and localization of discontinuities is an important extension that should be added to the methods described in this book.

2.5   Alternative representations

In this chapter, we have examined the use of visible surface representations for describing intrinsic images. Other computer vision researchers have used a variety of alternative representations, both in conjunction with low-level vision algorithms such as stereo matching and in higher level processing such as navigation or object recognition. In this section, we will briefly mention some of these alternatives, and compare their characteristics with those of the visible surface representation.

One of the most common variations on the single valued visible surface representation is the multiple valued disparity field. In this representation, a number of possible disparities d are represented at each retinotopic location (x, y). The three dimensional cube of data (x, y, d) can contain either binary values (representing depth hypotheses) or analog values (representing the confidence in these hypotheses). This representation has been used in many stereo correspondence algorithms (Marr and Poggio 1976, Drumheller and Poggio 1986), and is capable of representing transparent surfaces (Prazdny 1985). While it is easy to represent competing hypotheses during stereo matching, it is not obvious how to encode a physically meaningful measure of smoothness. A more general discussion of these considerations appears in (Szeliski 1986).

A variety of representations have been used for integrating range data obtained from a mobile robot. Spatial occupancy maps use a grid aligned with the floor to represent the "certainty" of each cell being occupied (Elfes and Matthies 1987, Moravec 1988). The grid can also be used to represent the elevation of terrain in an outdoor environment (Hebert and Kanade 1988, Hebert *et al.* 1988). Three-dimensional occupancy maps have been used successfully to obtain high resolution maps from sonar data (Stewart 1987). These grid-based representations are more suitable than visible surface representations for fusing information from a wide variety of viewpoints, especially since most surface interpolation algorithms are viewpoint dependent (Blake and Zisserman 1986a). However, it is difficult to tell from these representation where the object surface might lie, and to localize its position to less than the grid cell size.

The visible surface representation used in this book has been generalized to three-dimensional objects through the use of energy-based models (Terzopoulos *et al.* 1987). These symmetry-seeking models (or "air bags") use a finite element representation parameterized by cylindrical coordinates that encodes the three-dimensional position of each nodal variable. A similar model can be developed using an elastic net (Durbin and Willshaw 1987). The error modeling techniques developed in this book can be applied directly to these energy-based models, although Fourier analysis techniques are no longer applicable.

An alternative to energy-based surface representations is the use of spatial likelihood maps (Christ 1987). This method uses a probability density function defined over a spherical coordinate system to integrate location and surface normal information obtained from a tactile sensor. Like the disparity field and occupancy map representations, a single surface is not represented, and smoothness constraints are difficult to implement.

Another family of representations that is popular for object and part representation is that of lumped parameter models. These models, which include generalized cylinders (Brooks *et al.* 1979, Shafer and Kanade 1983) and superquadrics (Pentland 1986), define three-dimensional surfaces (and volumes) using a small number of global geometric parameters. These models are obviously at a higher level than the visible surfaces which we have been studying, but they can often be used for similar applications. The parameter values for these models can often be directly estimated from intensity or range data. The small dimensionality of these parameter sets makes their estimation more robust and can lead to easier matching at later stages. However, these models are less flexible in terms of representing a wide variety of shapes compared to the distributed models which use energy-based finite element representations.

While these alternative representations may offer advantages over visible surface representations in some situations, in this work we concentrate on the latter representation. Visible surface representations can be used to model a wide variety of intrinsic images and can be used to integrate the output of independent low-level vision modules. This intermediate representation can serve as a basis for higher level processes such as segmentation and grouping. In this chapter, we have seen how regularization, finite element analysis and relaxation can be used to formulate and solve low-level vision problems. We have shown how multiresolution representations can increase the efficiency of these algorithms and can also increase the descriptive power of the representation. Lastly, we have discussed how discontinuities play an important role in the development of a suitably flexible and accurate representation. In the next chapter, we will examine how a Bayesian formulation can be used in conjunction with these representations to develop more sophisticated low-level vision algorithms.

Chapter 3

Bayesian models and Markov Random Fields

In the early days of computer vision, Bayesian modeling was a popular technique for formulating estimation and pattern classification problems (Duda and Hart 1973). This probabilistic approach fell into disuse, however, as computer vision shifted its attention to the understanding of the physics of image formation and the solution of inverse problems. Bayesian modeling has had a recent resurgence, due in part to the increased sophistication available from Markov Random Field models, and due to a realization of the importance of sensor and error modeling. In this chapter, we will briefly review the general Bayesian modeling framework. This will be followed by an introduction to Markov Random Fields and their implementation. We will then discuss the utility of probabilistic models in later stages of vision and preview the use of Bayesian modeling in the remainder of the book.

3.1 Bayesian models

A Bayesian model is a statistical description of an estimation problem which consists of two separate components. The first component, the *prior model*, $p(\mathbf{u})$, is a probabilistic description of the world or its properties before any sense data is collected. The second component, the *sensor model*, $p(\mathbf{d}|\mathbf{u})$, is a description of the noisy or stochastic processes that relate the original (unknown) state $\mathbf{u}$ to the sampled input image or sensor values $\mathbf{d}$. These two probabilistic models can be combined to obtain a *posterior model*, $p(\mathbf{u}|\mathbf{d})$, which is a probabilistic description of the current estimate of $\mathbf{u}$ given the data $\mathbf{d}$. To compute this posterior model we use Bayes' Rule

$$p(\mathbf{u}|\mathbf{d}) = \frac{p(\mathbf{d}|\mathbf{u})\,p(\mathbf{u})}{p(\mathbf{d})} \tag{3.1}$$

where

$$p(\mathbf{d}) = \sum_{\mathbf{u}} p(\mathbf{d}|\mathbf{u}).$$

In its usual application (Geman and Geman 1984), Bayesian modeling is used to find the *Maximum A Posteriori* (MAP) estimate, i.e., the value of $\mathbf{u}$ which maximizes the conditional probability $p(\mathbf{u}|\mathbf{d})$. In the more general case (Section 6.1), the optimal estimator $\mathbf{u}^*$ can be the solution that minimizes the expected value of a loss function $L(\mathbf{u}, \mathbf{u}^*)$ with respect to this conditional probability. As we will show in Section 3.3, additional useful information (such as the uncertainty in our estimates) can be extracted from the posterior distribution.

A simple example of Bayesian modeling is the classification of terrain according to multispectral satellite data (this example is adapted from Duda and Hart (1973)). Consider the problem of classifying terrain into vegetation ($\star$) and non-vegetation ($\circ$) based on the readings from a single infrared band. The prior probability of the two classes at any pixel is

$$p(u_i = \star) = 0.7 \quad \text{and} \quad p(u_i = \circ) = 0.3$$

(in this example, all pixel values are uncorrelated). The two sensor model curves $p(d_i|u_i = \star)$ and $p(d_i|u_i = \circ)$ are shown in Figure 3.1a. By applying Bayes' Rule, we can compute the posterior distributions $p(u_i = \star|d_i)$ and $p(u_i = \circ|d_i)$ as shown in Figure 3.1b. As we can see from this figure, there exists a single threshold that can be used for classifying pixels from the sensor data. This restricted use of Bayesian modeling is called *Bayes decision theory* (Duda and Hart 1973).

To use the Bayesian framework in conjunction with visible surface representation, we must somehow encode the smoothness inherent in these fields. We can do this by using the prior model to describe the correlation between adjacent pixels. A simple method for modeling such correlation is presented next.

3.2   Markov Random Fields

A Markov Random Field is a probability distribution defined over a discrete field where the probability of a particular variable u_i depends only on a small number of its neighbors,

$$p(u_i|\mathbf{u}) = p(u_i|\{u_j\}), \quad j \in N_i. \tag{3.2}$$

We can use MRF's to model the correlated structure of dense fields or the smoothness inherent in visible surfaces. For our terrain classification example, we can specify conditional probabilities for particular configurations of terrain type (Figure 3.2). In this figure, we see that the existence of a particular terrain type ($\star$ or $\circ$) is less likely if it is surrounded by different terrain than if it has similar neighbors.

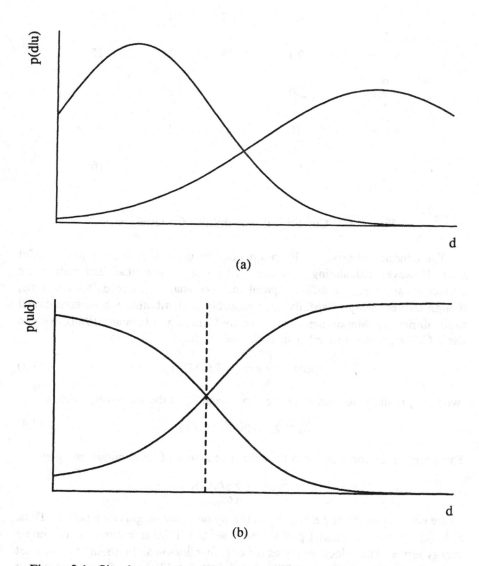

(a)

(b)

Figure 3.1: Simple example of Bayesian modeling: terrain classification

(a) sensor models (b) posterior estimates

Configuration	$p(u_i = \star\|\mathbf{u})$	$\dfrac{p(u_i=\star\|\mathbf{u})}{p(u_i=o\|\mathbf{u})}$	Energy	new $\dfrac{p(u_i=\star\|\mathbf{u})}{p(u_i=o\|\mathbf{u})}$
o o ⋆ o o	.05	.05	4	.06
o o ⋆ o ⋆	.20	.25	3	.25
o o ⋆ ⋆ ⋆	.50	1	2	1
o ⋆ ⋆ ⋆ ⋆	.80	4	1	4
⋆ ⋆ ⋆ ⋆ ⋆	.95	19	0	16

Figure 3.2: Conditional probabilities for terrain model

The conditional probabilities $p(u_i|\mathbf{u})$ can be used to generate a prior model $p(\mathbf{u})$. However, calculating $p(\mathbf{u})$ such that all of the marginal distributions are correct is in general a difficult problem. Fortunately, there exists a simple, though indirect, way of specifying a probability distribution whose conditional probabilities are Markovian. As shown by Geman and Geman (1984), we can use a Gibbs (or Boltzmann) distribution of the form

$$p(\mathbf{u}) = \frac{1}{Z_p} \exp(-E_p(\mathbf{u})/T_p), \tag{3.3}$$

where T_p is the *temperature* of the model and Z_p is the *partition function*

$$Z_p = \sum_{\mathbf{u}} \exp(-E_p(\mathbf{u})/T_p). \tag{3.4}$$

The energy function $E_p(\mathbf{u})$ can be written as a sum of local clique energies

$$E_p(\mathbf{u}) = \sum_{c \in C} E_c(\mathbf{u}),$$

where each clique energy $E_c(\mathbf{u})$ depends only on a few neighboring points. Thus, to build up our conditional probabilities, we use a linear summation of simple energy terms. These local energies (or cost functions) can be thought of as a set of *weak constraints* (Hinton 1977) that penalize unlikely configurations of our prior model.

For our terrain model example, we can use a very simple prior energy

$$E_p(\mathbf{u}) = \sum_{(i,j)}(1 - \delta(u_{i,j}, u_{i+1,j})) + (1 - \delta(u_{i,j}, u_{i,j+1})) \tag{3.5}$$

which simply counts the number of adjacent pixels which differ. Using this model, we can calculate the local energy $E_p(u_i|\mathbf{u})$ of the configurations in Figure 3.2, as shown in the fourth column. By computing the energy associated with the two values for the center pixel, we can derive the probability ratio of the two configurations

$$\frac{p(u_i = \star|\mathbf{u})}{p(u_i = \circ|\mathbf{u})} = \qquad (3.6)$$

Choosing $T_p = (\ln 2)^{-1}$, we obtain the values shown in the fifth column of Figure 3.2. These values are close to the ones that were originally specified. To obtain exactly the same values, higher order (three element) cliques would have to be used (Geman and Geman 1984).

To compute the probability of any configuration $\mathbf{u}$ using (3.3) is straightforward, but it may be prohibitively expensive due the exponential complexity of the partition function. For most applications, however, this computation is not necessary. If we wish to generate a random sample from the distribution (3.3), we can use an algorithm called the Gibbs Sampler (Geman and Geman 1984). This iterative algorithm successively updates each state variable u_i by randomly picking a value from the local Gibbs distribution

$$p(u_i|\mathbf{u}) = \frac{1}{Z_i} \exp(-E_p(u_i|\mathbf{u})/T_p) \qquad (3.7)$$

where

$$Z_i = \sum_{u_i} \exp(-E_p(u_i|\mathbf{u})/T_p).$$

This random updating rule is guaranteed to converge (in the ensemble sense) to a representative sample from the Gibbs distribution. To speed up this convergence, simulated annealing (Metropolis *et al.* 1953, Kirkpatrick *et al.* 1983, Hinton and Sejnowski 1983) can be used. The stochastic multigrid techniques discussed in Section 4.2 (and also Barnard (1989), Konrad and Dubois (1988)) can also be used to speed up convergence.

Applying the Gibbs Sampler to our simple terrain classification example, we find that a typical random sample looks like Figure 3.3a. Notice that the terrain appears in clumps. Even though the energy function used in the Gibbs distribution contains only local terms, the Markov Random Field itself has long-range interactions (i.e., even points that are far apart are still correlated). We can use this correlated prior model to improve our terrain classification algorithm. For our measurement model, we will use the truncated Gaussian distributions shown in Figure 3.1b

$$p(d_i|u_i = v) = \frac{1}{Z_v} \exp(-\frac{(d_i - \mu_v)^2}{2\sigma_v^2}) \qquad (3.8)$$

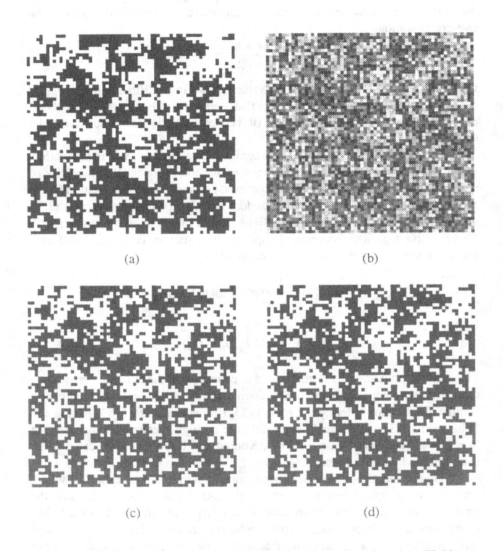

(a)   (b)

(c)   (d)

Figure 3.3: Restoration of noisy images through Markov Random Fields

(a) typical MRF (b) noisy sample (c) MAP estimate (d) MPM estimate

with

$$(\mu_0, \sigma_0) = (0.2, 0.2) \quad \text{and} \quad (\mu_1, \sigma_1) = (0.8, 0.3).$$

Applying this noisy measurement process to the random sample that we have just generated, we obtain the image shown in Figure 3.3b. For calculating the posterior distribution, it will be convenient to use the log probability of the measurement distribution

$$E_d^i(u_i, d_i) \equiv -\log p(d_i|u_i) = \sum_{v \in \{0,1\}} \delta(u_i, v) \left[\frac{(d_i - \mu_v)^2}{2\sigma_v^2} + \log Z_v \right]. \tag{3.9}$$

We can thus write the joint distribution of the measurement model as

$$p(\mathbf{d}|\mathbf{u}) = \frac{1}{Z_d} \exp(-E_d(\mathbf{u}, \mathbf{d})) \tag{3.10}$$

with

$$E_d(\mathbf{u}, \mathbf{d}) = \sum_i E_d^i(u_i, d_i).$$

We are now in a position to derive the posterior distribution $p(\mathbf{u}|\mathbf{d})$ using Bayes' Rule. From (3.1), (3.3) and (3.10) we have

$$p(\mathbf{u}|\mathbf{d}) = \frac{p(\mathbf{d}|\mathbf{u})p(\mathbf{u})}{p(\mathbf{d})} = \frac{1}{Z} \exp(-E(\mathbf{u})) \tag{3.11}$$

where

$$E(\mathbf{u}) = E_p(\mathbf{u})/T_p + E_d(\mathbf{u}, \mathbf{d}). \tag{3.12}$$

We thus see that the posterior distribution is itself a Markov Random Field. To compute the MAP estimate, we need only to minimize $E(\mathbf{u})$.

The energy function described by (3.12) has many local minima, so we must use simulated annealing to perform the optimization. The Gibbs Sampler algorithm (using $E(\mathbf{u})$ as the energy function) can be used directly to find the MAP estimate, so long as the system is frozen at the end of the annealing (Geman and Geman 1984). The result of applying the MAP algorithm to our sampled image is shown in Figure 3.3c. Alternatively, we could also calculate the *Maximizer of Posterior Marginals* (Marroquin 1985), which minimizes the expected number of misclassified pixels (Figure 3.3d).

Comparing (3.12) to the regularization equation (2.1) developed in the previous chapter, we see that regularization is an example of the more general Bayesian approach to optimal estimation. This observation has been made previously in both the numerical analysis literature (Kimeldorf and Wahba 1970) and in the computer vision field (Terzopoulos 1986b, Bertero *et al.* 1987). Some newly discovered implications of this relationship will be discussed in Section 4.1. The Bayesian interpretation of regularization will also be used in Chapter 6 to develop uncertainty estimation and parameter estimation techniques.

The example that we have used to explain Markov Random Fields is a simple version of the MRF approach to image restoration developed by Geman and Geman (1984) and Marroquin (1985). Markov Random Fields have also been used for solving the stereo correspondence problem (Marroquin 1985, Szeliski 1986, Barnard 1986) and for determining discontinuities in visible surfaces (Marroquin 1984). In this latter application, the line processes which we encountered in the previous chapter can be used to represent the discontinuities (Geman and Geman 1984). The use of line processes to encode and localize discontinuities is currently one of the chief attractions of the MRF approach to low-level vision (Poggio *et al.* 1988).

Despite their attractive computational properties and their flexibility, Markov Random Fields have some limitations. Markov Random Fields represent distributions with a particularly simple structure, and may be unsuited for modeling more complicated distributions or even distributions with limited correlations (see Section 4.1). MRFs are good at modeling fields or surfaces such as terrain maps that have a certain smoothness or coherence but that can have many bumps or wiggles. They are less appropriate for modeling surfaces with more global properties such as piecewise planar surfaces[1]. The direct estimation and modeling of global geometric parameters may be more appropriate in such cases (Durrant-Whyte 1987).

3.3   Using probabilistic models

The Bayesian models and Markov Random Fields that we have introduced in this chapter have previously been used to obtain optimal estimates such as those shown in Figure 3.3. In this book, we will argue that additional useful information can be extracted from the posterior distribution, and that a probabilistic development of prior and sensor models can yield new insights into the solution of low-level vision problems.

A simple way to make better use of a posterior distribution is to calculate higher order statistics (such as variance) to measure the uncertainty in our estimates. The variance of each point can be calculated independently as

$$\text{Var}(u_i) = \sigma_i^2 = \int (u_i - u_i^*)^2 p(\mathbf{u}|\mathbf{d})\, d\mathbf{u}. \tag{3.13}$$

The full covariance matrix of the field $\mathbf{u}$ can also be calculated as

$$\text{Cov}(\mathbf{u}) = \Sigma_{\mathbf{u}} = \int (\mathbf{u} - \mathbf{u}^*)(\mathbf{u} - \mathbf{u}^*)^T p(\mathbf{u}|\mathbf{d})\, d\mathbf{u}, \tag{3.14}$$

---

[1]A Markov Random Field can sometimes be designed such that it has the desired surface (e.g., a plane) in its null space (Leclerc 1989). However, the estimation of the surface using the MRF will be much slower than direct least squares fitting.

but this information may be too voluminous to store for reasonably sized fields. Higher order statistics could also be estimated, but these are not examined in this monograph for reasons of simplicity. In many cases, the distributions that we deal with will be multivariate Gaussians, so that the first and second order statistics completely capture the information about the distribution.

An alternative to calculating higher order statistics is to pass the whole probability distribution to the next higher level of processing. This information can then be used to answer specific questions, such as determining if a point belongs to the surface, or implementing certainty-weighted model matching. In general, a full posterior probability distribution may be too complex to describe. In the case of MRFs, however, the distribution can be specified with only a few parameters per pixel.

Maintaining a probabilistic description of our current estimate is particularly useful in the context of dynamic vision. In such a system, new information is continually being acquired due to either observer or scene motion, and estimates are continually being updated. A useful formalism for modeling such a system is the Kalman filter, which we will examine in Chapter 7. The Kalman filter approach extends the Bayesian framework introduced in this chapter by adding a *system model* to the prior and measurement models. This system model describes the evolution of the state being estimated and contains a stochastic component to account for unknown disturbances and model errors.

The work described in this book differs from traditional Bayesian modeling in that some of the parameters in the prior or sensor models may be unknown and may need to be estimated from the data. In the case of prior models, this may involve estimating the amount of smoothing to be applied (Section 6.3) or the shape of the smoothing (which is related to the correlation of the prior model). For sensor models, it may be necessary to determine the confidence in individual measurements from the data itself (Section 5.3).

The general Bayesian modeling framework presented in this chapter can be instantiated in many ways, depending on the particular visual task, visual domain, and sensing strategies being studied. In the next three chapters, we will examine, in turn, prior models, sensor models, and posterior models. The prior models that we will study are based on Markov Random Fields and regularization. The relationship of these prior models to fractal surfaces will be examined. We will also develop relative multiresolution representations by examining the spectral characteristics of the prior model. A variety of sensor models will then be developed, including sparse depth sensors with one-dimensional and three-dimensional Gaussian noise and an optical flow estimator. Probabilistic posterior models will then be developed. From these models, we will devise new techniques for estimating posterior uncertainty, estimating regularization parameters, and estimating observer motion. Finally, the Bayesian framework will be extended to handle temporal sequences by developing Kalman filter-based algorithms for on-line estimation in a dynamic environment.

Chapter 4

Prior models

As we have seen in the previous chapter, prior models play an essential role in the formulation of Bayesian estimators. A prior model can be as simple as the prior probabilities of different terrain types used in our remote sensing example of Section 3.1, or as complicated as the initial state (position, orientation and velocity) estimate of a satellite in a Kalman filter on-line estimation system. When applied to low-level vision, prior models encode the smoothness or coherence of the two-dimensional fields that are being estimated from the image. In this chapter, we will examine the spectral characteristics of our prior models, develop algorithms for efficiently generating random samples, develop a relative representation using a frequency domain approach, and compare our probabilistic models to deterministic (mechanical) models. Let us start by previewing how these four ideas fit together.

The use of Markov Random Fields for modeling smooth fields was first suggested by Geman and Geman (1984). In their implementation, they used discrete values for the intensity and an energy function similar to (3.5), which favored piecewise continuous surfaces. They were also the first to use line processes in conjunction with a MRF representation. Subsequent research has used fields whose energy resembles that obtained from discretizing the membrane model (Marroquin 1984). In Section 4.1 we will extend this idea by examining the effect of using the stabilizers used in regularization to define our probabilistic prior models. In particular, we will show how the choice of stabilizer determines the power spectrum of the prior model. We will also compare these MRF models to models based on assuming a particular correlation structure for the surface.

The ability to generate sample elements from our model space is one of the attractions of the probabilistic approach. This capability allows us to determine if these random samples are consistent with our intuitions about the domain that we are modeling. To generate these typical samples, we will use the Gibbs Sampler algorithm described in Section 3.2. As we will show in Section 4.2, the implementation of this algorithm for models such as the membrane and thin plate is particularly simple and only requires adding a controlled amount of Gaussian noise to the usual Gauss-Seidel relaxation algorithm. We will examine how multigrid (coarse-to-fine) stochastic relaxation can help speed up the approach of the Gibbs Sampler towards equilibrium. We will also show how this algorithm can be used for generating constrained fractals for computer graphics applications.

In Section 4.2, we re-examine the question of designing a multiresolution relative representation. Using the results of Section 4.1, we show how to design a multiresolution representation by matching the sum of the power spectra at each level to the spectrum of the original single-resolution implementation. This composite representation thus has a smoothing and interpolating behavior similar to that of the original model (since they implement similar priors), and also decomposes the signal into a multiresolution description.

The prior models that we will be studying in this chapter are commonly used to describe intrinsic images and can thus be thought of as "intrinsic models" (this term was coined by Gudrun Klinker). In the hierarchy of visual processing (Figure 2.1), intrinsic models span the middle ground between the object models used in high-level vision and the physical models that describe image formation. Object models are normally used to determine the identity and pose (position and orientation) of a three-dimensional object. These models are typically described by a small number of lumped parameters, such as the pose, the relative positions of parts for articulated objects, and perhaps some shape parameters for models such as superquadrics (Pentland 1986). In certain cases, the parameters of these models can be determined directly from the image data (Lowe 1985, Pentland 1986). Intrinsic models, on the other hand, have a large number of distributed parameters, such as the depth value at each node for a surface model. If we are to recover these parameters from the limited data available in the image, we must restrict their values and thus restrict the space of possible models.

Two approaches that can be used to achieve this restriction are energy-based models (e.g., regularization) and Bayesian prior models. As we discussed in the previous chapter, these two approaches are somewhat equivalent. In their recent book, Blake and Zisserman (1987) argue that the mechanical (energy-based) approach is preferable to the probabilistic (Bayesian) one. In Section 4.4, we will give counterarguments in support of the Bayesian approach, and show how our method can be extended to include weak continuity constraints and three-dimensional models.

4.1　Regularization and fractal priors

When selecting a Markov Random Field prior model, we must choose some energy function to define the Gibbs distribution. As we saw in Section 3.2,

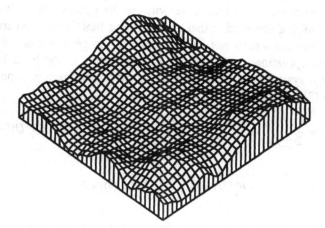

Figure 4.1: Typical sample from the thin plate prior model

choosing the regularization smoothness constraint as the energy function results in a MAP estimate which is identical to that obtained from regularization. While this observation has been used as a statistical justification for regularization, the characteristics of the prior model have not previously been investigated. In this section, we will study the prior model in isolation and determine what class of surfaces it describes.

One way of studying the prior model is to generate some typical random samples using the Gibbs Sampler algorithm described in Section 3.2 (the implementation details are given in the next section). Using the thin plate whose energy is given in (2.4) as our model, we can generate a typical sample from the prior distribution as shown in Figure 4.1. This surface has an interesting rough or bumpy structure that is quite different from the smooth shape that one might expect. A convenient way to characterize this roughness is to compute the spectral characteristics of the surface using Fourier analysis.

The Fourier transform (Bracewell 1978) of a multidimensional signal $v(\mathbf{x})$ is defined by

$$\mathcal{F}\{v\} \equiv \int v(\mathbf{x}) \exp(2\pi i \mathbf{f} \cdot \mathbf{x}) \, d\mathbf{x} = V(\mathbf{f}), \qquad (4.1)$$

and the transform of its partial derivative is given by

$$\mathcal{F}\{\frac{\partial v(\mathbf{x})}{\partial x_j}\} = (2\pi i f_j) \, V(\mathbf{f}). \qquad (4.2)$$

Using Rayleigh's energy theorem

$$\int |v(\mathbf{x})|^2 d\mathbf{x} = \int |V(\mathbf{f})|^2 d\mathbf{f}, \qquad (4.3)$$

we can re-write the smoothness functional $E_p(\mathbf{u})$ in terms of the Fourier transform $U(\mathbf{f}) = \mathcal{F}\{u\}$ to obtain the new energy function $E'_p(U)$.

For our smoothness functional, we will use the general form given in (2.6) with the simplifying assumption that the weighting functions $w_m(\mathbf{x})$ are constant. While this assumption does not strictly apply to the general case of piecewise continuous interpolation, it provides an approximation to the local behavior of the regularized system away from boundaries and discontinuities. Applying (4.2) and (4.3) to (2.6) we obtain

$$E'_p(U) = \frac{1}{2} \sum_{m=0}^{p} \int w_m \sum_{j_1 + \cdots + j_d = m} \frac{m!}{j_1! \cdots j_d!} \left| (2\pi i f_1)^{j_1} \cdots (2\pi i f_d)^{j_d} U(\mathbf{f}) \right|^2 d\mathbf{f}$$

or

$$E'_p(U) = \frac{1}{2} \int |H_p(\mathbf{f})|^2 |U(\mathbf{f})|^2 d\mathbf{f} \qquad (4.4)$$

where

$$|H_p(\mathbf{f})|^2 = \sum_{m=0}^{p} w_m |2\pi \mathbf{f}|^{2m}. \qquad (4.5)$$

For the membrane interpolator, $|H_p(\mathbf{f})|^2 \propto |2\pi \mathbf{f}|^2$ and for the thin plate model, $|H_p(\mathbf{f})|^2 \propto |2\pi \mathbf{f}|^4$.

The results of this analysis can be combined with a similar analysis of the data constraint to derive the filtering behavior of regularization-based smoothers (see Appendix B). In this section, however, we are interested in studying the spectral characteristics of the prior model. To derive these, we note that since the Fourier transform is a linear operation, if $u(\mathbf{x})$ is a random variable with a Gibbs distribution with energy $E_p(\mathbf{u})$, then $U(\mathbf{f})$ is a random variable with a Gibbs distribution[1] with energy $E'_p(U)$. We thus have

$$p(U) \propto \exp\left(-\frac{1}{2} \int |H_p(\mathbf{f})|^2 |U(\mathbf{f})|^2 d\mathbf{f} \right)$$

from which we see that the probability distribution at any frequency $\mathbf{f}$ is

$$p(U(\mathbf{f})) \propto \exp\left(-\frac{1}{2} |H_p(\mathbf{f})|^2 |U(\mathbf{f})|^2 \right).$$

Thus, $U(\mathbf{f})$ is a random Gaussian variable with variance $|H_p(\mathbf{f})|^{-2}$, and the signal $u(\mathbf{x})$ is correlated Gaussian noise with a spectral distribution

$$S_u(\mathbf{f}) = |H_p(\mathbf{f})|^{-2}. \qquad (4.6)$$

From this analysis, we can conclude that using a regularization-based smoothness constraint is equivalent to using a correlated Gaussian field as the Bayesian

---

[1]This is because the Jacobian $|\partial U/\partial u|$ is a constant for a linear operator.

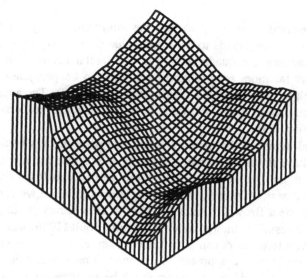

Figure 4.2: Fractal (random) solution

prior. The spectral characteristics of this Gaussian field are determined by the choice of stabilizer. For the membrane and the thin plate models, we have

$$S_{\text{membrane}}(\mathbf{f}) \propto |2\pi\mathbf{f}|^{-2} \tag{4.7}$$

and

$$S_{\text{thin-plate}}(\mathbf{f}) \propto |2\pi\mathbf{f}|^{-4}. \tag{4.8}$$

These equations are interesting because they correspond in form to the spectra of Brownian fractals.

Fractals are a class of mathematical objects that exhibit self-similarity over a range of scales (Mandelbrot 1982). Fractals have been used to generate intricate geometric designs, to study the statistical properties of coastlines and structured noise, and to generate realistic images of terrain. A stochastic fractal is a random process or a random field that exhibits self-affine statistics over a range of scales. A common way to characterize such a fractal is to say that it follows a power law in its spectral density

$$S_v(f) \propto 1/f^{\beta}. \tag{4.9}$$

This spectral density characterizes a fractal Brownian function $v_H(\mathbf{x})$ with $2H = \beta - E$, whose fractal dimension is $D = E + 1 - H$ (where E is the dimension of the Euclidean space) (Voss 1985). A function that satisfies (4.9) may also be fractional Gaussian noise (Rensink 1986).

Comparing (4.7) or (4.8) to (4.9), we can conclude that the smoothness assumptions embedded in certain regularization methods are equivalent to assuming that the underlying processes is fractal (Szeliski 1987). When regularization techniques are used, it is usual to find the minimum energy solution (Figure 2.2c), which also corresponds to the mean value solution for those cases where the energy functions are quadratic. Thus, the fractal nature of the process is not evident. A far more *representative* solution can be generated if a random (fractal) sample is taken from this distribution (Figure 4.2). The amount of noise (and hence roughness) that is desirable or appropriate can be derived from the data (see Section 6.3).

The fractal nature of the membrane and thin plate models means that these interpolators have no natural scale, and can thus be applied to any size of data[2]. It also suggests that we could use priors with in-between (truly fractional) degrees of smoothness. In theory, this is straightforward, since we can specify the prior model to be a Brownian fractal field with an arbitrary β. In practice, implementing the resulting interpolator is difficult. Boult (1986) has implemented such fractional interpolators using reproducing kernel splines. In Section 4.2 we will present an alternative approach based on multiresolution relaxation.

The prior models which we use need not be isotropic or homogeneous. In general, we can choose a prior model with any arbitrary correlation function[3] (the correlation is the inverse Fourier transform of the spectral density). We could thus model mountain ridges or terrain with varying degrees of smoothness within a single framework. We will explore how to implement such arbitrary prior models in the next section.

Since the prior models which we have examined in this section are examples of correlated Gaussian noise, it is interesting to compare these surfaces to those obtained by filtering white noise through a finite impulse response (FIR) filter. A regularization-based prior model described by (4.5) and (4.6) is a pseudo-Markovian Gaussian field of order p (Adler 1981). This field has an infinite correlation, which is what makes it particularly suitable for interpolation. The discrete version of this field is a Markov Random Field, and its energy function can be described by a small number of additive terms[4]. The generative model for this field is the Gibbs Sampler, which can be quite slow since the system must be near equilibrium to generate a representative sample. The MAP estimate can be found using a similar relaxation algorithm.

In contrast, FIR filtered white noise has a finite correlation and is non-Markovian (no two sample points are conditionally independent)[5]. The generative model for this field is extremely simple (just filter white noise). However, local relaxation algorithms for doing MAP estimation do not in general exist. We can thus see the two main advantages of using Markov Random Fields for

---

[2]This scale independence is in marked contrast to the viewpoint dependence of these models (Blake and Zisserman 1986a).

[3]This suggestion was made by Alex Pentland.

[4]In fact, its information (inverse covariance) matrix is sparse and banded (see 6.2).

[5]The exception is white noise, which is both an MRF and (trivially) FIR filtered white noise.

modeling surfaces and other intrinsic images: they can model infinite range correlations with only local interactions, and they lead to local relaxation algorithms for MAP estimation.

4.2   Generating constrained fractals

To generate the random samples from either the prior or posterior models, we have to use the Gibbs Sampler algorithm described in Section 3.2. As we have seen, the ability to generate such random samples from the prior model gives us a useful tool for studying the behavior of the prior. In Section 6.2, we will see how this random sampler can also be used to determine the uncertainty in the posterior estimate. In this section, we will describe our implementation of the Gibbs Sampler for visible surfaces, and show how this new algorithm can be used to generate constrained fractal surfaces for computer graphics applications.

As we saw in Section 3.2, the Gibbs Sampler is an iterative stochastic algorithm where each state variable u_i is updated asynchronously (sequentially) by picking a value from the local Gibbs distribution (3.7). For the visible surface representations that we studied in Section 2.2, the local energy function (2.17) is quadratic, with a minimum value u_i^+ given by (2.18) and a second derivative equal to a_{ii}. The local Gibbs distribution is therefore

$$p(u_i|\mathbf{u}) \propto \exp\left(-\frac{a_{ii}(u_i - u_i^+)^2}{2T_p}\right) \qquad (4.10)$$

which is a Gaussian with mean u_i^+ and variance T_p/a_{ii}. We thus see that the Gibbs Sampler is equivalent to the usual Gauss-Seidel relaxation algorithm with the addition of some locally controlled Gaussian noise at each step (Szeliski 1987). The temperature parameter T_p controls the amount of roughness in the random sample. In Section 6.3, we will present a method for determining the appropriate value of T_p from the sampled data.

As is the case with deterministic relaxation, the above algorithm converges very slowly towards its equilibrium distribution (the point at which the system exhibits negligible statistical dependence on its starting configuration (Ackley *et al.* 1985)). To speed up this convergence, we can use a coarse-to-fine technique similar to the one used with deterministic relaxation. We simply generate a random sample using the Gibbs Sampler at a coarser level, and then use the interpolated sample as a starting configuration for the finer level. This starting configuration will already be closer to equilibrium than a non-random configuration such as the zero state. More importantly, it will contain more of the low-frequency components of the random field than can be obtained by iterating for a long time on the fine level. Appendix B contains a Fourier analysis of the convergence rates of the multiresolution Gibbs Sampler. Multiresolution stochastic resolution has also been studied by Barnard (1989) and Konrad and Dubois (1988).

For a correct implementation of the multiresolution algorithm, we must ensure that the coarse level generates a random sample which has the same statistics as the subsampled version of a random fine level sample. In general, this is a difficult problem which may require the application of group renormalization techniques (Wilson 1979). Fortunately, when the energy equations are quadratic, it suffices to ensure that the energy equations at the coarse level are a good approximation to the energy of the *deterministic* continuous system (which can be ensured by using finite element analysis). To prove that this is sufficient, we need to show that the distribution obtained from substituting the *minimum* energy configuration that matches the coarse level sample into the fine level energy equations is close to the true probability distribution.

Consider a fine level system u_1 with an associated energy equation

$$E_1(\mathbf{u}_1) = \frac{1}{2}(\mathbf{u}_1 - \hat{\mathbf{u}}_1)^T \mathbf{A}_1(\mathbf{u}_1 - \hat{\mathbf{u}}_1). \qquad (4.11)$$

This equation implies that $\mathbf{u}_1$ is a multivariate Gaussian with mean $\hat{\mathbf{u}}_1$ and a covariance $\mathbf{A}_1^{-1}$. We can obtain a coarse level system $\mathbf{u}_2$ using a subsampling matrix $\mathbf{H}$, i.e., $\mathbf{u}_2 = \mathbf{H}\mathbf{u}_1$. The coarse level system is thus a multivariate Gaussian with mean $\hat{\mathbf{u}}_2 = \mathbf{H}\hat{\mathbf{u}}_1$ and a covariance $\mathbf{H}^T \mathbf{A}_1^{-1} \mathbf{H}$. Let $\bar{\mathbf{u}}_1$ denote the minimum energy solution to (4.11) which satisfies $\mathbf{u}_2 = \mathbf{H}\bar{\mathbf{u}}_1$. We wish to show that the coarse level distribution is the same as would be obtained by substituting $\bar{\mathbf{u}}_1$ into (4.11). To do this, we consider the joint distribution $p(\mathbf{u}_1, \mathbf{u}_2)$, which is also a multivariate Gaussian. Using the result shown in Appendix D, we can show that marginalizing a multivariate Gaussian with respect to some of its state variables is equivalent to substituting the minimum energy solution for those variables (in terms of the remaining variables) into the joint density function. Since in our case $\bar{\mathbf{u}}_1 = \mathbf{H}\mathbf{u}_2$, the joint density is equivalent to the prior distribution on $\mathbf{u}_1$, and we thus have the desired result

$$p_2(\mathbf{u}_2) = p_1(\bar{\mathbf{u}}_1). \qquad (4.12)$$

The significance of this result is as follows. In general, to derive a coarse level probability density function from a fine level function, we must marginalize over all of the possible configurations of the fine model that are consistent with the coarse model. For the case of our visible surface models, whose typical elements are rough fractals, we expect the average energy of a fine surface coincident with the coarse level solution to be much higher than the minimum energy possible. Fortunately, this difference in energy is a constant independent of the coarse level configuration. Since we are using a Gibbs distribution, this means that the probability density obtained using the minimum energy (smooth) solution is the same as would be obtained by taking into account the rough fractal nature of the prior model.

By considering the continuous surface model as the fine level, this same argument can be used to justify why it is acceptable to develop finite element energy equations assuming that the system is always in its minimum energy state, and to then use these energy equations in the probability distribution. This unique situation only exists when the distributions are Gaussian fields (or equivalently, the energy functionals are quadratic). In the general case, such as for spin glasses or other systems with an underlying discrete structure and non-quadratic energy functions, we have to use more sophisticated techniques (Wilson 1979, Barnard 1989). The same is true for our visible surface representations when we include line processes as part of the state description (rather than keeping them fixed). In our current implementation, we ignore the problems associated with the re-normalization of line process variables and simply use energy-based approximations.

An example of our multiresolution Gibbs Sampler being used to generate a fractal surface is shown in Figure 4.3 (this example uses a thin plate energy model). The solution at each level is used as a starting point for the Gibbs Sampler at the next finer level. Since iteration at the finer level can significantly move the solution away from its starting point, the coarse levels are no longer subsampled versions of the fine level. If we wish the fine level solutions to have a closer resemblance to the coarse level solution (say for generating a zoom sequence in computer graphics (Fournier *et al.* 1982)), we can either use fewer iterations on the fine level, or add an extra data compatibility constraint between the coarse and fine level surfaces (Szeliski and Terzopoulos 1989a).

We can use the multiresolution Gibbs Sampler algorithm which we have just developed to generate constrained fractals with arbitrary discontinuities. Using the same data points as we used for the thin plate interpolation example (Figure 2.2a) and also the same energy equations, we can apply the Gibbs Sampler to the posterior distribution defined by (3.11). A typical sample generated by this approach is shown in Figure 4.2. While this sample is not truly fractal since it depends on the data points, it is a typical sample from the fractal prior distribution conditioned on the data points that were observed. We can thus shape the fractal by imposing arbitrary depth constraints, orientation constraints (Terzopoulos 1984), depth discontinuities or creases.

A number of different fractal generation algorithms have been developed over the last few years for computer graphics applications. Fournier *et al.* (1982) use a technique called random midpoint displacement which creates a tessellated surface by recursively subdividing triangles. As new interior points are created, their height value is randomly perturbed away from their original interpolated value, with the magnitude of the perturbation being related to the level of the subdivision. By varying the relationship between this perturbation

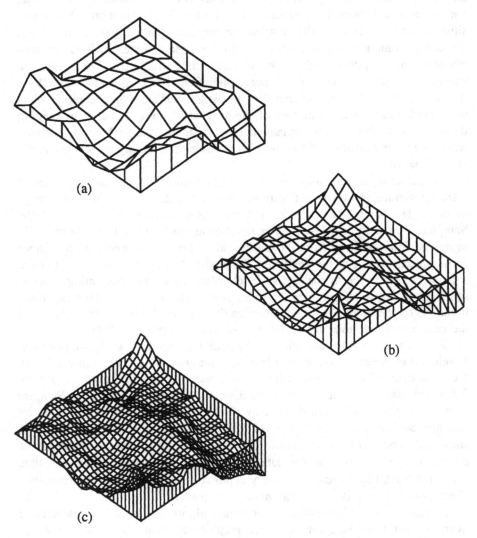

(a)

(b)

(c)

Figure 4.3: Multiresolution fractal sample from thin plate model

magnitude and the subdivision level, fractals of arbitrary degree can be created. Voss (1985) uses successive random additions, which is similar to random midpoint displacement, except that *all* of the points are randomly perturbed at each subdivision step (not just the ones which are newly created). Recently, Lewis (1987) has introduced a refinement of random midpoint displacement which he calls generalized stochastic subdivision. Instead of displacing each midpoint independently, correlated Gaussian noise is added. This reduces the creases that are sometimes visible with the former method.

All three of these methods can implement fractals of arbitrary dimension, but suffer from statistics that are not spatially stationary. Their biggest deficit is that they are difficult to constrain. The usual approach to this problem is to first lay out a coarse solution, and to then add fractal texture to this solution using recursive subdivision. In contrast to these previous methods, our multiresolution Gibbs Sampler easily accommodates arbitrary constraints as additional terms in its energy equation. The solution is also guaranteed to have spatially stationary statistics, so long as the fine level solution is iterated a sufficient number of times (in practice, this number can be quite low). The biggest limitation of the multiresolution Gibbs Sampler as we have thus far described it is that it is limited to prior models with spectra of the form

$$S_u(\mathbf{f}) \propto |\mathbf{f}|^{-2m},$$

i.e., surfaces that have integral fractal dimensions. We thus have to extend our algorithm to overcome this limitation.

The simplest way to approximate a fractal with an in-between fractal dimension is to use a blend of the thin plate and membrane models (Terzopoulos (1986b) calls these "splines under tension"). If we choose $w_1 = |2\pi f_0|^2 w_2$ in (2.6), we obtain the power spectrum shown in Figure 4.4. This Bode plot (log-log power plot) shows that the prior model behaves as $S_u \propto |\mathbf{f}|^{-3}$ in the vicinity of f_0 (2.0 in this case), as a membrane at lower frequencies, and as a thin plate at higher frequencies. The interpolated solution using this mixed model is shown in Figure 4.5a, and a typical sample from the posterior distribution is shown in Figure 4.5b.

We can extend the range of in-between fractal behavior by modifying the coarse-to-fine Gibbs Sampler. Instead of implementing the same energy equation at each level, we modify the algorithm so that a different blend of membrane and thin plate is used at each level. Since we wish to have $S_u \approx |\mathbf{f}|^{-\beta}$, we require that

$$|H_p^{l+1}(\mathbf{f})|^2 = 2^{-\beta}|H_p^l(2\mathbf{f})|^2 \tag{4.13}$$

where $|H_p^l(\mathbf{f})|^{-2}$ is the power spectrum of the prior model at level l. From this equation, we obtain the relationship

$$w_m^{l+1} = 2^{2m-\beta}w_m^l. \tag{4.14}$$

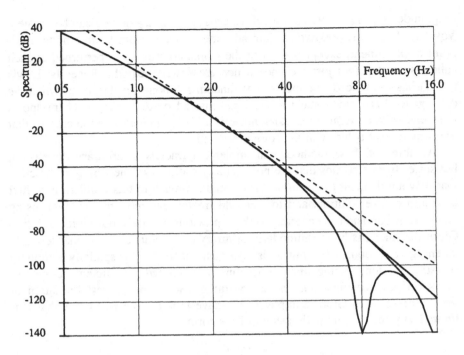

Figure 4.4: Power spectrum of mixed membrane / thin plate
The coarse, medium and fine level spectra are shown (the fine level curve is uppermost). The dashed line shows the $|f|^{-3}$ asymptote.

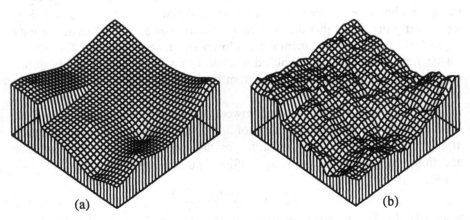

Figure 4.5: Interpolated surface and fractal surface for mixed membrane / thin plate

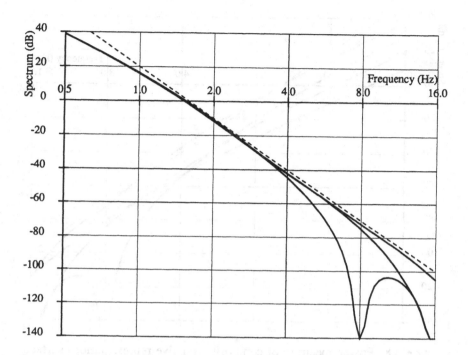

Figure 4.6: Power spectrum of fractal approximation

The coarse, medium and fine level spectra are shown (the fine level curve is uppermost). The dashed line shows the $|f|^{-3}$ asymptote.

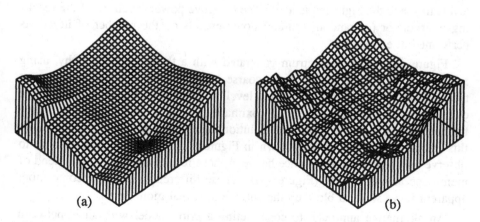

Figure 4.7: Interpolated surface and fractal surface for fractal approximation

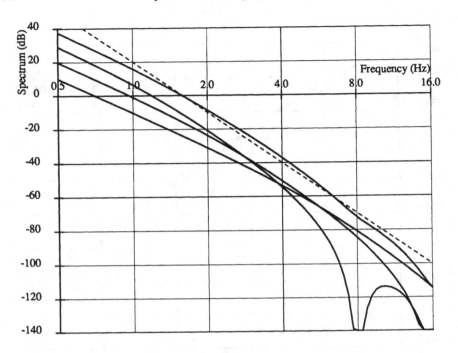

Figure 4.8: Power spectrum of composite (relative representation) surface
The coarse, medium and fine level spectra are shown, along with the summed
spectrum (uppermost solid curve). The dashed line shows the $|f|^{-3}$ asymptote.

Since the relaxation at a given level mostly affects the high frequency components, we can use the coarse levels to shape the low frequencies and the fine
levels to shape the high frequencies. The effective power spectrum of the resulting interpolator (or random sample) now depends on the number of iterations
performed at each level.

Figure 4.6 shows a spectrum generated with a three level hierarchy, using
a large number of iterations at the coarse level followed by five iterations at
the medium level and five at the fine level. Compared to Figure 4.4, the range
over which the spectrum closely approximates $|f|^{-3}$ is extended from one octave
to three octaves. The interpolated solution and typical posterior sample using
this fractal approximation are shown in Figures 4.7a and 4.7b. It is difficult to
observe any significant difference between these results and the simple blend of
membrane and thin plate (Figure 4.5). These differences might become more
apparent if we were to blow up the surface to reveal more detail.

An alternative approach to constructing a prior model with an in-between
fractal spectrum is to create a composite fractal by summing up the interpolated
depth maps from each level in a multiresolution hierarchy. Using this approach,

each level has its own associated prior energy function, and the power spectrum of the composite surface is the sum of the individual level spectra (since we are adding independent Gaussian fields). To ensure that the summed spectrum has the desired fractal dimension, we impose the same condition on the w_m values at each level as we used with the coarse-to-fine Gibbs Sampler (4.14).

The resulting composite power spectrum for a three level pyramid is shown in Figure 4.8. The shape of this spectrum is not as linear as that shown in Figure 4.6, but this situation could be improved by using more coarse levels. The main advantage of using the composite fractal is that the spectrum is not dependent on the number of iterations performed at each level (more precisely, each level must be iterated until it is at equilibrium). Additional advantages include the ability to perform relaxation at all levels simultaneously and a decomposition of the signal into a multiresolution relative representation[6]. This method, however, is sensitive to the choice of interpolator used at each level and may also have very slow convergence towards equilibrium because of the tight coupling between levels when doing interpolation (MAP estimation). We will defer these problems until the next section, where we re-examine the multiresolution relative representation in light of our new results about fractal priors.

Before closing our discussion of fractal generation, it should be pointed out that we can easily choose non-homogeneous or non-isotropic fractal models[7]. In fact, we can choose arbitrary correlations for our prior model, although these may lead to energy equations whose discrete implementation does not have a sparse (MRF) structure. Figure 4.10 shows an example of the sophisticated fractals that can be generated using our new multiresolution Gibbs Sampler method. This fractal scene is constrained by the data points shown in Figure 4.9, and has a crease coinciding with the ravine on the upper right side, and a depth discontinuity on the lower right. Four different continuity models are used: a blend of membrane and thin plate in the upper quadrant, a membrane in the right quadrant, a thin plate in the left quadrant, and a very stiff thin plate in the lower quadrant.

The methods which we have described in this section can be improved in several ways. A better representation for discontinuities should lead to more natural looking breaks, and a more sophisticated implementation of the finite element discretization could increase the power of the method (e.g., by allowing the placement of data points on non-lattice locations). The use of weak continuity constraints could also be investigated, allowing our fractal generation algorithm to introduce tears into the surface automatically. Finally more efficient algorithms for attaining equilibrium should be investigated.

---

[6]In fact, we can use this approach to invert the fractal process, i.e., to decompose a fractal surface into its multiresolution description.

[7]The issue of non-isotropic interpolators has also been studied by Boult (1986).

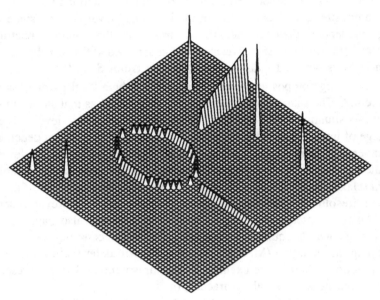

Figure 4.9: Depth constraints for fractal generation example

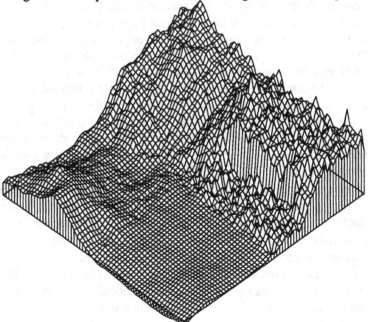

Figure 4.10: Constrained fractal with spatially varying fractal dimension and variance

In summary, the new fractal generation algorithm which we have developed in this section has several advantages over existing techniques. It allows depth and orientation constraints as well as depth and orientation discontinuities to be imposed on the fractal generation process at arbitrary locations. The resulting fractal has homogeneous statistics (conditioned on the constraints), and these statistics can be varied locally. The coarse-to-fine algorithm which we developed efficiently produces a random sample from this constrained fractal, which can be of arbitrary degree or spectral shape.

4.3   Relative depth representations (reprise)

In Section 2.3, we saw how a multiresolution relative representation could decompose the visible surface into a number of local depth maps, each of which would describe surface detail at its own level of resolution. Discontinuities could be assigned to just one level in the resolution hierarchy, and each level could also have a separate degree of uncertainty, thus permitting the representation of fine detail even when the absolute depth is unknown. Such a multiresolution representation could also use fully parallel relaxation.

Our investigation into relative representations was hampered by the difficulty of selecting appropriate prior models and interpolation functions for the levels in the hierarchy. We derived an equation (2.25) which placed conditions on the prior information matrices A_p^l and the interpolation matrices I_l which would ensure equivalence to a single-resolution model. Unfortunately, this equation is not very intuitive and involves inverse information matrices, which are not sparse.

A simpler development can be obtained using the results presented in this chapter. Instead of defining the prior model by a smoothness energy functional, we consider the prior model to be a correlated Gaussian field with a known power spectrum. We can then calculate the power spectrum of the composite (summed) surface in the relative representation by adding up the spectra of the levels multiplied by the filtering behavior of the interpolators. This design in frequency space is significantly simpler than trying to manipulate large matrices.

To formalize this description, let $|H_p^l(\mathbf{f})|^{-2}$ be the spectrum of the prior model at level l. From Appendix B, we see that

$$|H_p^l(\mathbf{f})|^2 = \sum_{m=0}^{p} w_m h_l^{-2m} (4 - 2\cos 2\pi f_x h_l - 2\cos 2\pi f_y h_l)^{2m}$$

where $h_l = |\Delta x_l| = |\Delta y_l|$ is the grid spacing at level l. Alternatively it can be obtained by taking the Fourier transform of a row of the spline model matrix A_p^l,

$$|H_p^l(\mathbf{f})|^2 = |\mathcal{F}\{[A_p^l]_i\}|^{-1}. \tag{4.15}$$

This power spectrum closely approximates (4.5) for frequencies that are small compared to $1/h_l$ (i.e., the approximation error of the finite element implementation is proportional to the grid size). Let $F^l(f)$ be the frequency response of the interpolation filter. For example, for the bilinear interpolator, we have

$$F^l(f) = \left(\frac{\sin \pi f_x h_l}{\pi f_x h_l}\right)^2 \left(\frac{\sin \pi f_y h_l}{\pi f_y h_l}\right)^2.$$

The composite spectrum can now be derived as

$$|\tilde{H}_p(f)|^{-2} = \sum_{l=1}^{L} |F^l(f)|^2 |H_p^l(f)|^{-2}. \tag{4.16}$$

Comparing this equation to (2.25) (Section 2.3), we see that the new equation deals with spectral characteristics, which are easy to visualize and for which established design techniques exist. For example, if we wish to use our relative representation to model surfaces with a given fractal spectrum, we can use the relationship given in (4.14), which results in the spectrum shown in Figure 4.8. Figure 4.11 shows the five-level relative representation with a fractal degree $\beta = 3$ applied to our sample data set. For this example, a blend of a membrane and a thin plate is used at each level along with third-degree Langrange interpolation.

Some care must be taken with the spectral approach to prior modeling. Since we wish the composite surface, which is a Gaussian field, to have spatially homogeneous (shift-invariant) statistics, we have to ensure that the fields interpolated from each level also have this characteristic. This condition is equivalent to ensuring that the Fourier transform of interpolated field is uncorrelated noise, or equivalently, that the power in the sidebands (the region $|f| > 1/h_l$) is small compared to the power at the other (finer) levels. In practice, this means that we have to use an interpolator of a sufficiently high degree to avoid visible creases or other artifacts that arise from interpolating the coarse level.

The initial results with the relative representations are encouraging, but much work remains to be done. The biggest current problem is with the speed of convergence. Because of the tight coupling between the levels, usual Gauss-Seidel relaxation does not work well. Conjugate gradient descent (Press *et al.* 1986) does better, since it can quickly get out of the narrow ravines in the energy landscape that are caused by the tight coupling. Another possibility worth exploring is to initially relax the data coupling term, since this again makes the system less stiff (Hinton 1977). Using higher order interpolators should also produce better (smoother) results, although at the expense of longer computation times. Finally, the introduction of discontinuities at the appropriate level, and the interaction between discontinuities at different levels, should prove to be a promising area of research.

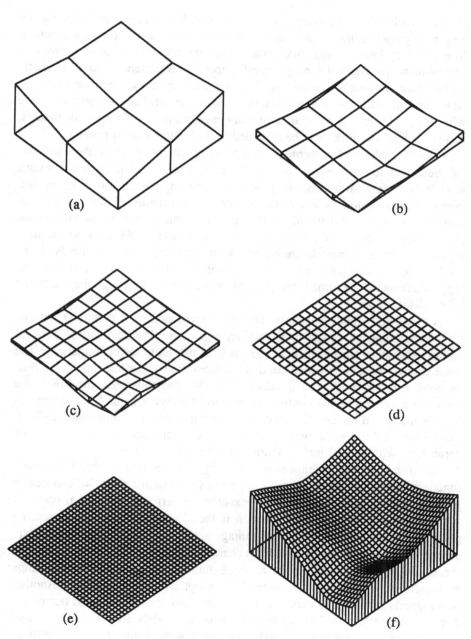

(a)

(b)

(c)

(d)

(e)

(f)

Figure 4.11: Relative representation for $\beta = 3$ interpolator

4.4   Mechanical vs. probabilistic models

In their book *Visual Reconstruction*, Blake and Zisserman (1987) lay out an elegant approach to the problem of piecewise continuous surface reconstruction. In explaining their method, Blake and Zisserman argue in favor of adopting a deterministic mechanical (energy-based) approach in preference to a stochastic probabilistic approach. In this section, we will argue the converse, that the Bayesian approach has many advantages over the mechanical approach. We will also study the use of weak continuity constraints in defining prior models, and show how our ideas can be extended to three-dimensional models.

One of Blake and Zisserman's main arguments in favor of the mechanical viewpoint is that the models should be continuous. Using such continuous models facilitates variational analysis, allows the implementation of viewpoint-invariant interpolators, and matches the continuous nature of surfaces and intensity fields in the real world. Fortunately, continuous models are not incompatible with probabilistic modeling. As we have seen in the previous section, regularization-based models are equivalent to correlated Gaussian fields. Even models such as viewpoint-invariant interpolators that do not have quadratic energy functions can be turned into probability distributions through the use of the Gibbs distribution.

A continuous field, of course, cannot be simulated on a computer without first discretizing the energy or probability equations. The best way to perform this discretization for mechanical models is to use finite element analysis. The same discrete equations which are derived from the mechanical model can then be used to define a Markov Random Field. We can thus view the MRF as a discretized version of the continuous pseudo-Markovian field. The parameters for this field need not be computed by assigning conditional probabilities. They can be derived the same way as they are for mechanical models, i.e., using parameters with natural interpretations or physical correlates.

Because of the equivalence between energy functions and probability density functions which can be established using the Gibbs distribution, we can design probabilistic models which have the exact same performance as that obtained with mechanical models. The question is then "what possible, advantages do probabilistic models offer?" One advantage is that we can develop probabilistic sensor models which closely match the characteristics of real sensors. As we will see in Chapter 5, we can model the three-dimensional noise inherent in range measurements or analyze the uncertainty in optical flow estimation. Another advantage is that we can choose different loss functions to use with our posterior estimator (Section 6.1), whereas mechanical models always find the MAP estimate. With a probabilistic model we can also determine the uncertainty in our estimate (Section 6.2), which corresponds to determining the stiffness of the mechanical model. Additional applications of the probabilistic approach will be presented in Chapters 6 and 7.

Perhaps the most interesting result of Bayesian modeling, however, is what it tells us about the continuity of the models that we use in low-level vision. A thin plate spline is piecewise continuous in its third derivative (therefore continuous everywhere in its second), with the discontinuities coinciding with the data points. A typical random sample from the thin plate prior model, on the other hand, is discontinuous everywhere in its second derivative (in fact, its second derivative is white noise). The minimum energy solution is thus smoother that an actual *typical* sample from which the data may have come; this smoothness is attributable to the averaging behavior of the mean estimate. We thus have an interesting dichotomy: the minimum energy solution is smoother but shows higher order discontinuities coincident with the data points, whereas a typical sample from the posterior distribution is rougher but shows no clues as to where the data points might be located.

So far, we have been discussing prior models that have a uniform continuity across the visual field. These models give rise to fractal surfaces which may be appropriate for modeling terrain or uninterpreted image data, but which seem like poor models for most visual surfaces. The weak continuity models used by Blake and Zisserman (1987) seem much more appropriate for the structured environments in which we live. To investigate if these weak continuity models are indeed useful as a prior model, we can use their energy equations as part of a Gibbs distribution and generate typical samples using the Gibbs Sampler.

Figure 4.12a shows a random sample taken from the weak membrane prior model. Compared to the regular membrane model shown in Figure 4.12c, the weak membrane has less roughness in the continuous areas since we can run this model at a lower temperature. The depth discontinuities, whose locations are shown in Figure 4.12b, allow for a greater variation in depth than would be possible with the continuous membrane. By turning down the temperature even lower while decreasing the line-break penalty, we can obtain a system which models piecewise constant surfaces and whose behavior resembles the model used by Leclerc (1989) in his scene segmentation work.

Using weak thin plates instead of weak membranes should lead to better prior models and better posterior estimates. With an appropriate selection of parameters, we can use these to model piecewise planar surfaces. Even higher order models are possible. For example, we could use a third order Tikhonov stabilizer ((2.6) with non-zero w_3). Blake and Zisserman (1987) have also suggested using models which are piecewise concave or piecewise convex, which corresponds to a weak continuity constraint on the sign of the curvature. Unfortunately, as the complexity of the model increases, so does the time required to generate a typical sample or the time to find a MAP estimate. Nevertheless, these possibilities deserve further study.

An alternative to using weak continuity constraints or other energy-based

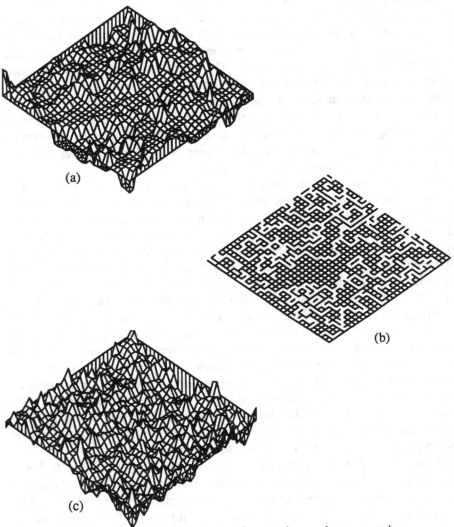

Figure 4.12: Random sample using weak membrane as prior

(a) random sample (b) discontinuities shown as *missing* lines (c) using regular membrane

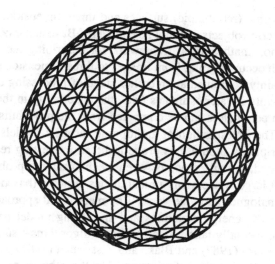

Figure 4.13: Random sample from a three-dimensional elastic net

models is to use process models to define the priors. As discussed by Mandelbrot (1982), process models describe repeated actions such as breaking or bending operating at a number of resolutions whose compound action results in a fractal surface. For the fractal surfaces examined in the previous section, the process model is rather simple. We take an elastic sheet and bombard it at every point with random vertical disturbances (white Gaussian noise) (Szeliski and Terzopoulos 1989a). More sophisticated process models which introduce rips and creases into the surface or bend the surface by applying torques at sparse locations should result in prior models that better describe visible surfaces and other intrinsic images.

Probabilistic modeling also need not be restricted to two-dimensional fields. We can apply the same techniques for converting energy-based models into Bayesian priors to three-dimensional models such as those being investigated by Terzopoulos *et al.* (1987). Figure 4.13 shows a three dimensional elastic net (based on the two-dimensional nets developed by Durbin and Willshaw (1987)) which was obtained by tessellating a sphere. The energy equation for this net is

$$E(\mathbf{p}_i) = \sum_i \sum_{j \in N_i} |\mathbf{p}_i - \mathbf{p}_j|^2 - \rho \sum_i |\mathbf{p}_i|.$$

Applying the Gibbs Sampler to this system, we obtain the typical sample shown in Figure 4.13. This figure resembles the examples of fractal textured spheres shown in (Mandelbrot 1982) and (Pentland 1986) which were generated by adding fractal texture onto the surface of a sphere.

From this example, we see that the difference between intrinsic models that describe visible (retinotopic) surfaces and three-dimensional energy-based models that describe objects may not be that large. Bayesian modeling may thus serve as a common mathematical framework for describing the multiple transformations which occur in going from images to three-dimensional models. Moreover, the uncertainty computed at an earlier stage of processing can be used to derive the uncertainty in later estimates (e.g., the uncertainty in the object model shape or position parameters can be computed from the probabilistic description of the surface). The investigation of appropriate intrinsic models for the intermediate level descriptions should thus prove to be an interesting research topic.

In the end, the difference between mechanical and probabilistic models may not be that large, since many probabilistics systems (based on MRFs) have mechanical analogues and vice versa. The mechanical approach may be preferable for developing energy equations and specifying model parameters. For MAP estimation, specially tailored deterministic algorithms—such as those developed by Witkin *et al.* (1987) and Blake and Zisserman (1987)—should perform better than general stochastic optimization. On the other hand, the probabilistic approach enables the development of more sophisticated estimates, including the use of different loss functions and the estimation of model parameters. We will examine these and other advantages of the probabilistic approach in the next three chapters of this book.

Chapter 5

Sensor models

Modeling the error inherent in sensors and using these error models to improve performance are becoming increasingly important in computer vision (Matthies and Shafer 1987). In the context of the Bayesian estimation framework, sensor models form the second major component of our Bayesian model. In this chapter, we will examine a number of different sensor models which arise from both sparse (symbolic) and dense (iconic) measurements.

One of the most common measurements used in building visible surface representations is the sparse depth measurement, which can be obtained either by passive methods such as stereo triangulation or by active methods such as laser range finding. In Section 5.1, we will examine how the uncertainty associated with such a depth measurement can be modeled by the stiffness of a spring attached to the surface. In Section 5.2, we will show how the full three-dimensional uncertainty of the measurement can be incorporated into a sensor model. We then turn our attention to dense measurements. In Section 5.3, we analyze the uncertainty inherent in a correlation-based flow measurement algorithm. Lastly, in Section 5.4, we present a simple imaging model of a CCD camera.

The four sensor models presented in this chapter are just a few of the many sensor models which could be developed. Using techniques similar to the ones presented here, we could develop error models for shape from shading, spatio-temporal velocity filters, and laser range finders, to mention a few. By choosing just a few examples, we aim to demonstrate in this chapter the advantages of sensor modeling, and to support the usefulness of the Bayesian modeling framework.

5.1   Sparse data: spring models

A noisy depth measurement, such as the three-dimensional location of a feature obtained by stereo triangulation, can be characterized by a three-dimensional probability distribution. Although the shape of this distribution may be quite complex, it can often be approximated by a 3-D Gaussian (Matthies and Shafer 1987). An advantage of using a Gaussian is that the position vector $\mathbf{p}_i = (x_i, y_i, z_i)$ and the 3×3 covariance matrix $\mathbf{C}_i$ completely specify the distribution.

To determine the interaction between a data point and the visible surface which we are building from our depth measurements, we must first convert this three-dimensional distribution in space into a one-dimensional distribution in depth. Assuming that x and y are the underlying natural coordinates of our visible surface representation, we set $d_i = z_i$ and use the probability distribution

$$p(d_i|\mathbf{u}) = \frac{1}{\sqrt{2\pi}\sigma_{z_i}} \exp\left(-\frac{(u(x_i, y_i) - d_i)^2}{2\sigma_{z_i}^2}\right). \tag{5.1}$$

If the errors for the various depth measurements are uncorrelated, which is usually the case, this leads to the usual data compatibility equation (2.2). When our retinotopic representation is not aligned with the sensor reference frame, the position and covariance measurements must first be transformed using simple matrix algebra (Matthies and Shafer 1987).

The effect of these depth constraints on the visible surface is similar to tying the surface to the depth values through springs (Terzopoulos 1984). The strength of the spring constants is inversely proportional to the variance of the depth measurements. To convert the analog constraints (5.1) to a discrete form, we usually associate the constraint with the nearest node. This introduces a quantization error; Terzopoulos (1984) compensates for this error by scaling the spring constant by an h^{-2} factor (where h is the grid spacing).

An alternative way to derive the discrete form of the data constraint is to write out the measurement equation

$$d_i = \mathbf{H}\mathbf{u} + r_i, \quad \text{where} \quad r_i \sim N(0, \sigma_i^2). \tag{5.2}$$

The measurement matrix $\mathbf{H}$ encodes how the surface point $u(x_i, y_i)$ which gives rise to the measurement d_i is obtained from the nodal variables $\mathbf{u}$; the $\mathbf{H}$ matrix thus depends on the choice of interpolator. If we choose block (constant) interpolation, then the discrete data constraint equation is as before (i.e., we associate the depth constraint with the nearest nodal variable). If we choose bilinear interpolation, we have

$$u(x, y) = h_{00}u_{i,j} + h_{01}u_{i,j+1} + h_{10}u_{i+1,j} + h_{11}u_{i+1,j+1}$$

where $u_{i,j}, \ldots, u_{i+1,j+1}$ are the four nodal variables nearest to (x, y) and $h_{00}, \ldots, h_{11}$ are interpolation constants which depend on x and y. The data constraint equation

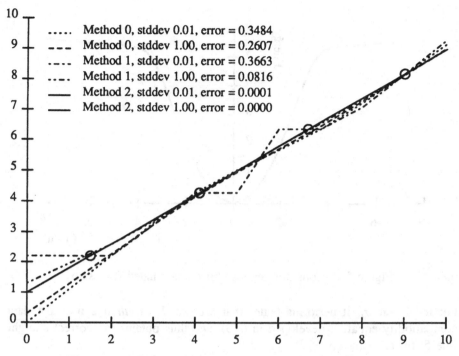

Figure 5.1: Cubic spline fit with different data constraints

thus becomes

$$E_d(d, \mathbf{u}) = \frac{1}{2}\sigma^{-2}\left(d - h_{00}u_{i,j} - h_{01}u_{i,j+1} - h_{10}u_{i+1,j} - h_{11}u_{i+1,j+1}\right)^2. \qquad (5.3)$$

This unfortunately introduces off-diagonal terms into our data compatibility matrix $\mathbf{A_d}$, and complicates the implementation of the finite element relaxation. We could drop the off-diagonal terms from $\mathbf{A_d}$, but this might introduce flat spots into our interpolated surface.

To explore the effects of these various choices on the quality of the surface fit, we can try a simple synthetic example. Figure 5.1 shows a number of cubic splines (the one-dimensional versions of a thin plate) which we fit to a set of four data points (shown as circles) which were sampled from a straight line. For each of the three data constraint methods described below, we fit the cubic spline to the data (using two different values of σ_i), and calculate the root mean squared (RMS) error between the fit and the original line. *Method 0* rounds the data points to the nearest grid point, and performs better when weaker springs (with a larger σ_i) are used. *Method 1* spreads the data constraints across the two neighboring nodes, without adding off-diagonal terms into the information matrix $\mathbf{A}$. For small σ_i, this approach creates marked flat spots in the solution,

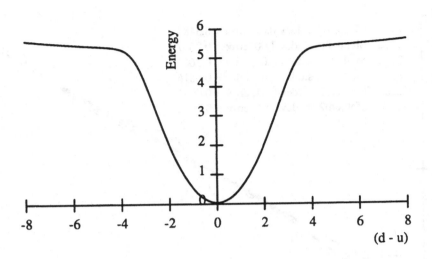

Figure 5.2: Constraint energy for contaminated Gaussian

but for larger σ_i, it performs better that *Method 1*. *Method 2* uses the data compatibility equation developed in (5.3), and thus obtains the correct straight line fit for all values of σ_i.

The probability distribution used to characterize the uncertainty in our depth measurement need not be Gaussian. The advantage of using a Gaussian is that the resulting constraint energy is quadratic. It is therefore easy to manipulate, and is characterized completely by its mean and variance values (first and second order statistics). A Gaussian distribution is appropriate when the error in the measurement is the result of the aggregation of many small random disturbances. Many sensors, however, have a normal operating range characterized by a small σ^2, but also occasionally produce gross errors. A more appropriate model for such a sensor is the *contaminated Gaussian* used by Durrant-Whyte (1987), which has the form

$$p(d_i|\mathbf{u}) = \frac{1-\epsilon}{\sqrt{2\pi}\sigma_1} \exp\left(-\frac{(u(x_i, y_i) - d_i)^2}{2\sigma_1^2}\right) + \frac{\epsilon}{\sqrt{2\pi}\sigma_2} \exp\left(-\frac{(u(x_i, y_i) - d_i)^2}{2\sigma_2^2}\right)$$

(5.4)

with $\sigma_2^2 \gg \sigma_1^2$ and $0.05 < \epsilon < 0.1$. This model behaves as a sensor with small variance σ_1^2 most of the time, but occasionally generates a measurement with a large variance σ_2. By taking the negative logarithm of the probability density function, we can obtain the constraint energy shown in Figure 5.2. This energy is similar in shape to the weak springs which arise in the weak continuity models of Blake and Zisserman (1987) and the ϕ function of Geman and McClure (1987).

Gaussians and contaminated Gaussians are just two of the many possible distributions that can be used to characterize sensors. The advantage of using a Bayesian approach to visual processing is that any sensor model that we develop can be incorporated directly into the estimation algorithm. In practice, finding good sensor models involves a tradeoff between the fidelity of the model, the compactness of its representation, and the tractability of its equations. We will see more examples of this tradeoff in the next section, where we examine how to better model the three-dimensional uncertainty inherent in sparse depth measurements.

5.2 Sparse data: force field models

One of the biggest deficiencies of the spring model that we have just presented is that it can only model the uncertainty in position along one dimension. In practice, most depth sensors have a full three-dimensional uncertainty associated with each position estimate (Matthies and Shafer 1987). Moreover, since the spring model ties the data point to exactly one attachment point on the surface, it is difficult to use with three dimensional models and parametric surfaces. In this section, we will introduce a new depth constraint model which incorporates three-dimensional uncertainty in point locations and thus overcomes these problems.

The key observation in developing such a model is that a sensed point can come from anywhere on the surface (within the limits of the sensor uncertainty). For example, if the sensed point $\mathbf{p}_i = (x_i, y_i, z_i)$ has a Gaussian distribution characterized by a covariance $\mathbf{C}_i$, then the likelihood that this measurement resulted from a noisy sensing of the surface point $\mathbf{q} = (x, y, u(x, y))$ is

$$p(\mathbf{p}_i|\mathbf{q}) = \frac{1}{\sqrt[3]{2\pi|\mathbf{C}_i|}} \exp\left(-\frac{1}{2}[\mathbf{p}_i - \mathbf{q}]^T \mathbf{C}_i^{-1}[\mathbf{p}_i - \mathbf{q}]\right).$$

Thus, to compute the likelihood of sensing $\mathbf{p}_i$ given a particular surface $u(x, y)$, we must integrate over all possible sites from which the point may have come,

$$p(\mathbf{p}_i|\mathbf{u}) = \int \frac{1}{\sqrt[3]{2\pi|\mathbf{C}_i|}} \exp\left(-\frac{1}{2}[\mathbf{p}_i - \mathbf{q}(x, y)]^T \mathbf{C}_i^{-1}[\mathbf{p}_i - \mathbf{q}(x, y)]\right) dx\,dy. \quad (5.5)$$

The constraint energy corresponding to this probability density function is a complicated non-linear function of the data and state variables. To study its properties, we will start with a few simple cases.

The simplest possible case is when we assume that the data came from one of the *nodal points*, i.e., from one of the points $\mathbf{q}_j = (x_j, y_j, u_j)$ where (x_j, y_j) are the Cartesian coordinates of the nodal variable u_j. In this case, we can replace the integral in (5.5) with a discrete summation

$$p(\mathbf{p}_i|\mathbf{u}) = \sum_j g_j(u_j) \quad (5.6)$$

where

$$g_j(u_j) = \frac{1}{\sqrt[3]{2\pi}|\mathbf{C}_i|} \exp{-f_j(u_j)}$$

and

$$f_j(u_j) = \frac{1}{2}[\mathbf{p}_i - \mathbf{q}_j]^T \mathbf{C}_i^{-1}[\mathbf{p}_i - \mathbf{q}_j].$$

The corresponding constraint energy equation is thus

$$E_{ff}(\mathbf{p}_i, \mathbf{u}) = -\log \sum_j g_j(u_j) \qquad (5.7)$$

which, unlike the models studied in the previous section, cannot be partitioned into a sum of local energies involving the data point $\mathbf{p}_i$ and the nodal variables u_j.

The energy equation (5.7) is interesting since it acts like a force field, attracting nearby surface points towards the data point, rather than tying the data to a particular fixed location. When the surface is intrinsically parameterized, i.e., when the three-dimensional surface point locations are functions of some intrinsic parameters (Terzopoulos *et al.* 1987), the energy equation behaves like a "slippery spring,"[1] allowing the surface to slide by the data point. This force field, derived from statistical considerations, is also qualitatively similar to the repulsive "volcano" used by Kass *et al.* (1988).

An energy equation similar to (5.7) has been used by Durbin and Willshaw (1987) in their algorithm for solving the Traveling Salesman Problem (their energy equation was derived heuristically, without any statistical interpretation). The form of the energy allows a small elastic net similar to a rubber band to be progressively stretched until it passes through all of the cities on the salesman's tour. To find the minimum energy solution, Durbin and Willshaw start with a large value of $\mathbf{C}_i$ (identical and spherically symmetric at each point), and slowly reduce its value while relaxing the net. Durbin *et al.* (1989) provide an analysis of the net behavior and its relationship to Bayesian modeling.

To gain some further insight into the behavior of our new energy constraint, we can calculate the instantaneous force that it exerts on the surface points. The force exerted on variable u_j can be calculated by differentiating (5.7) with respect to u_j

$$F_j = -\frac{\partial E_{ff}}{\partial u_j} = \frac{1}{\sum_k g_k(u_k)} \frac{\partial g_j}{\partial u_j} = \frac{g_j(u_j)}{\sum_k g_k(u_k)} \left(-\frac{\partial f_j}{\partial u_j}\right). \qquad (5.8)$$

The first part of this force equation, $g_j(u_j)/(\sum_k g_k(u_k))$, determines how the force from the data point is distributed among the surface points u_k. Since $g_k(u_k)$ corresponds to a Gaussian distribution, the magnitude of the force falls off like

---

[1]This term was suggested by Alex Pentland.

a Gaussian (skewed, of course, by the shape of the covariance matrix C_i). Note that there is a complex interaction between all of the points near p_i. If some other point u_k moves closer to p_i, the effective force at u_j will be reduced (similar observations have been made by Durbin and Willshaw (1987)).

The second part of the force equation (5.8) is

$$-\frac{\partial f_j}{\partial u_j} = a_{zz}(x_i - x_j) + a_{yz}(y_i - y_j) + a_{zz}(z_i - u_j) \quad \text{where} \quad C_i^{-1} = \begin{bmatrix} a_{xx} & a_{xy} & a_{xz} \\ a_{yx} & a_{yy} & a_{yz} \\ a_{zx} & a_{zy} & a_{zz} \end{bmatrix}.$$

If the measurement noise is not skewed (i.e., if C_i is diagonal), this force is simply proportional to $\sigma_{z_i}^{-2}(z_i - u_j)$, which is the same as the simple spring model used in the previous section. If the measurement noise is skewed (non-zero a_{xz} or a_{yz}), the null point of the spring (ignoring the first part of the energy term) can be calculated as

$$d_j = z_i + \frac{a_{xz}(x_i - x_j) + a_{yz}(y_i - y_j)}{a_{zz}}.$$

The interaction between the surface points and a single data point with 3-D uncertainty is illustrated in Figure 5.3. We can thus think of this model as implementing a *distributed spring* constraint.

The non-linear energy equation given by (5.7) can be used as part of a surface interpolation or other low-level vision algorithm, and models the three-dimensional uncertainty associated with the sparse depth measurement, unlike the simple spring model presented in the previous section. Assuming that the sensor noise is independent for each depth measurement, we can sum (5.7) over the points $\{p_i\}$ to obtain the composite energy equation $E_d(u, d)$ (here d stands for all of the data $\{p_i\}$). To solve the resulting set of regularized equations, we can use non-linear gradient descent techniques such as the ones used by Terzopoulos (1987) (see also Press *et al.* (1986)).

In many applications, however, we cannot afford to keep all of the data points p_i and their associated covariance matrices C_i. This is especially true in dynamic vision applications, where a single scene or object may be observed from many viewpoints. We thus have two choices for maintaining a constant amount of data while acquiring more measurements. The first approach, *symbolic matching*, finds a correspondence between the points $\{p_i^t\}$ seen at time t and the points $\{p_i^{t+1}\}$ seen at the next interval. The second approach, *iconic modeling*, approximates the complicated energy equation arising from the interaction of points described by (5.7) with a simple equation of the form used in standard regularization (2.13).

Symbolic matching is appropriate when the same surface points can be measured from different viewing directions or tracked over time. This is usually the case when the depth measurements result from matching discrete features

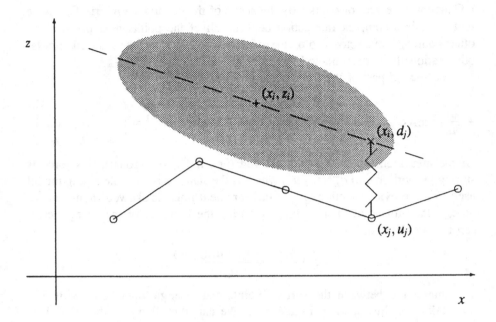

Figure 5.3: Depth constraint with three-dimensional uncertainty
The effect of data point (x_i, d_i) on nodal variable u_j is shown as a spring connecting u_j to the null position d_j. The strength of the spring depends on the location of the other neighboring points u_k and the shape of the noise covariance (indicated by a gray ellipse).

using stereo or motion (Matthies and Shafer 1987). Once the correspondence is known, we can compute the new updated estimates for the point location and covariance using the equations given in (Matthies and Shafer 1987) or (Matthies *et al.* 1987). After a series of n measurement, the variance in the point location is reduced by a factor of n in all directions[2].

The problem with symbolic matching is that it is often difficult to establish point correspondences. Moreover, certain depth sensors such as laser range finders do not even measure the same points when their viewing position changes (we will have more to say about this in Section 6.4). In this case, the iconic approach is more suitable. The iconic approach also has the advantage that the representations and associated algorithms are uniform and parallel, and thus map naturally onto massively parallel architectures.

To incorporate the data constraint (5.7) into an iconic framework (where all

---

[2]If the shape of the covariance matrix changes with viewing direction, an even greater reduction can be obtained.

quantities are represented as two-dimensional fields), we must approximate this equation with one of the form

$$E_d(\mathbf{p}_i, \mathbf{u}) = \frac{1}{2} \sum_j c_j (u_j - d_j)^2.$$

(5.9)

We should choose the c_j so that they encode the strength or stiffness of the constraint between u_j and d_j. Alternatively, we can view c_j as encoding the inverse variance (or information) of the constraint.

To compute the values for c_j, we can choose either to match the energy equations (5.7) and (5.9), or to match the probability densities from which these were derived. In this section, we will examine the former approach. A simple way to choose a value for c_j is to compute the second partial derivative of E_{ff} with respect to u_j

$$c_j = \frac{\partial^2 E_{ff}}{\partial u_j^2} = \frac{g_j(u_j)}{\sum_k g_k(u_k)} \left(\frac{\partial^2 f_j}{\partial u_j^2} + \left(1 - \frac{g_j(u_j)}{\sum_k g_k(u_k)} \right) g_j(u_j) \left(\frac{\partial f_j}{\partial u_j} \right)^2 \right).$$

(5.10)

We can ignore the second term in this expression if $\partial f_j / \partial u_j \approx 0$, which is true in the vicinity of the minimum energy configuration $u_j \approx d_j$. In this case, the formula for c_j reduces to

$$c_j = \frac{g_j(u_j)}{\sum_k g_k(u_k)} a_{zz}.$$

(5.11)

We can think of this formula as distributing the depth information available from the measurement at point $\mathbf{p}_i$ among the nodal variables u_j according to the ratio $g_j(u_j)/\sum_k g_k(u_k)$, (i.e., with a Gaussian fall-off). Note that the sum of all the local information quantities c_j is a_{zz}, the amount of depth information available in the original measurement (here we are using the term information interchangeably with inverse variance).

An alternative but closely related way to derive a tractable approximation to (5.7) is to use a Taylor series expansion around the current solution point $\mathbf{u}^*$,

$$E_d(\mathbf{p}_i, \mathbf{u}) = E_{ff}(\mathbf{u}^*) + \sum_j \left. \frac{\partial E_{ff}}{\partial u_j} \right|_{\mathbf{u}^*} (u_j - u_j^*) + \sum_j \sum_k \frac{1}{2} \left. \frac{\partial^2 E_{ff}}{\partial u_j \partial u_k} \right|_{\mathbf{u}^*} (u_j - u_j^*)(u_k - u_k^*).$$

(5.12)

Note that we only keep the first three terms in the Taylor series expansion. This allows us to write the data constraint energy as a quadratic form

$$E_d(\mathbf{p}_i, \mathbf{u}) = \frac{1}{2} \mathbf{u}^T A_i \mathbf{u} - \mathbf{b}_i^T \mathbf{u} + k_i$$

and to thus incorporate this constraint into our previously developed surface interpolation algorithms. We can use higher order terms to measure the skew in the local distribution, and to check the validity of our approximation. Expanding the solution around $\mathbf{u}^*$ instead of $\mathbf{d}$ (the minimum energy solution of E_{ff}) means that we either have to use the full form of $E_{ff}(\mathbf{p}_i, \mathbf{u})$ to find the solution, or that we have to apply (5.12) iteratively until we converge to such a solution. The advantage of expanding around $\mathbf{u}^*$ is that we can better approximate the behavior of our system near its current operating point.

This approach can in general be applied to any arbitrary constraint energy E_{ff}. If the energy equation is too complex to be differentiated analytically, the partial derivatives used in (5.12) can be evaluated numerically using small perturbations in the solution $\mathbf{u}^*$. We can also choose to ignore the off-diagonal (mixed partial derivatives) in (5.12), or to only evaluate the terms over limited spatial neighborhoods (since these terms denote the cross-correlation between adjacent nodal variables). The success of using (5.12) to model our sensor depends on the smoothness of the local energy landscape and how well it can be locally approximated by a quadratic form. We thus expect the method to work well with multivariate Gaussians, but perhaps not as well with uniform bounded distributions. The viability of our approach remains to be tested in practice.

When first developing (5.7), we assumed that the sensed data point came from one of the nodal points. This assumption was made mostly for simplicity of exposition. In practice, we can evaluate the integral equation (5.5) by choosing an interpolation function which maps from the nodal variables $\mathbf{u}$ to the surface $u(x, y)$. For a three-dimensional Gaussian distribution and a polynomial-based interpolator, the energy equation (and also its partial derivatives) can be analytically evaluated. The resulting equations involve Gaussians and error functions (integrals of Gaussians) and are thus quite cumbersome to manipulate. However, they yield a better model for the interaction between the data and the surface, especially when the point variance is small compared to the finite element grid size. In the limit, as the x and y uncertainties approach 0, the behavior of the three-dimensional uncertainty model approaches that of the one-dimensional model studied in the previous section.

When the iconic framework is used to integrate information over time, the variance in the depth estimate is reduced in proportion to the number of measurements acquired (just as with symbolic matching). The uncertainty in the (x, y) position, which is characterized by the distribution function $g_j(u_j) / \sum_k g_k(u_k)$, does not diminish. Thus, symbolic matching is preferable whenever possible, since the depth constraint becomes more localized. In those situations where such matching is impossible, however, the iconic approach developed in this section provides a tractable model of the three-dimensional uncertainty in sparse depth measurements.

The new sensor model which we have developed in this section has several interesting characteristics not found in previous models. Our model uses the full three-dimensional covariance of the sensor noise, and can easily be extended to model any three-dimensional probability distribution. It should therefore be useful in situations where the shape of the uncertainty is highly skewed, such as when range data is projected from camera coordinates to terrain coordinates (Hebert and Kanade 1988, Hebert *et al.* 1988). The constraints resulting from this sensor model do not tie the data points to any fixed points on the surface and are thus ideally suited for full three-dimensional models or parametric surfaces. The behavior of this data constraint can be similar to a force field or to a slippery spring, depending on the shape and size of the position uncertainty.

In addition to introducing the new sensor model, we have shown how it can be incorporated into our visible surface representation framework using a locally quadratic approximation to the energy function. This allows many measurements to be integrated over time while keeping the representation uniform and parallel. We will use this technique of developing locally quadratic approximations to the energy, which is equivalent to approximating the probability density by a multivariate Gaussian, at later points in the book.

5.3 Dense data: optical flow

Probabilistic sensor modeling need not be restricted to sparse measurements obtained directly from sensors. We can also apply error analysis to low-level vision algorithms and characterize these algorithms as *virtual sensors* with their own associated error models. In this section, we will analyze an intensity-based optical flow estimator and show how the uncertainty in the flow measurement at each point can be determined from local measurements already present in the algorithm. The flow field computed in this manner can be used as the input to a depth-from-motion algorithm (Section 7.2). The optical flow algorithm is an example of a virtual sensor that provides *dense* measurements; these measurements are similar in form to the visible surface representation into which they are integrated.

The problem of extracting optical flow from a sequence of intensity images has been extensively studied in computer vision. Early approaches used the ratio of the spatial and temporal image derivatives (Horn and Schunck 1981), while more recent approaches have used correlation between images (Anandan 1984) or spatio-temporal filtering (Heeger 1987). In this section, we analyze a simple version of correlation-based matching. This technique, called the Sum of Squared Differences (SSD) method by Anandan (1984), integrates the squared intensity difference between two shifted images over a small area to obtain an error measure

$$e_t(\mathbf{d}; \mathbf{x}) = \int w(\lambda)[f_t(\mathbf{x} + \mathbf{d} + \lambda) - f_{t-1}(\mathbf{x} + \lambda)]^2 \, d\lambda, \tag{5.13}$$

where f_t and f_{t-1} are the two intensity images, $\mathbf{x}$ is the image position, $\mathbf{d}$ is the displacement (flow) vector, and $w(\lambda)$ is a windowing function. The SSD measure is computed at each pixel **x** for a number of possible flow values **d**. In Anandan's algorithm, a coarse-to-fine technique is used to limit the range of possible flow values. In Matthies *et al.* (1987), a single-resolution algorithm is used since the range of possible motions is small due to high temporal sampling rates.

The resulting error surface $e_t(\mathbf{d}; \mathbf{x})$ is used to determine the best displacement estimate $\hat{\mathbf{d}}$ and the confidence in this estimate. Anandan and Weiss (1985) observed that the shape of the error surface differs depending on whether both, one or none of the displacement components are known (corresponding to an intensity corner, an edge, or a homogeneous area). They proposed an algorithm for computing the confidence measures based on the principal curvatures and the directions of the principal axes in the vicinity of the error surface minimum. Matthies *et al.* (1987) have shown how for a one-dimensional displacement, the variance in the displacement estimate can be computed from the second derivative in a parabola fit to the error curve (Lucas (1984) also has a similar analysis of gradient-based stereo matching).

In this section, we extend the result of Matthies *et al.* (1987) to two dimensions, thus providing a statistical justification for the heuristics developed by Anandan and Weiss. We start by fitting a quadratic of the form

$$e'_t(\mathbf{d}; \mathbf{x}) = (\mathbf{d} - \hat{\mathbf{d}})^T A (\mathbf{d} - \hat{\mathbf{d}}) + c \qquad (5.14)$$

to the error surface computed by (5.13). We then set the disparity estimate at **x** to $\hat{\mathbf{d}}$, and set the variance of this measurement to $2\sigma_n^2 A^{-1}$, where σ_n^2 is the variance of the white noise process. The inverse covariance matrix A can be singular, since it need not be inverted when being used in regularization-based smoothing of the flow field (Anandan and Weiss 1985).

The statistical justification for the above algorithm is given in Appendix C. The derivation involves modeling the two images frames f_t and f_{t-1} as displaced versions of the same image corrupted with additive white Gaussian noise. This simple model does not account for occlusions, disparity gradients or other optical effects. It is thus only valid over small windows[3], and breaks down in certain areas such as at occlusion boundaries. Nevertheless, it has proven to be remarkably effective in extract depth from small-motion sequences (Matthies *et al.* 1987). The analysis presented in Appendix C can also be used to derive the correlation between adjacent flow estimates and between flow estimates obtained from successive frames. These correlations are presently ignored in the algorithm developed by Matthies *et al.* (1987), but they could easily be integrated into an on-line estimation framework (Gelb 1974).

From the results presented in this section, we see how the statistical analysis of an optical flow algorithm can provide an error model for its output. This

---

[3]Note how the variance of estimate is reduced as the window size is increased, but the validity of the model becomes more suspect.

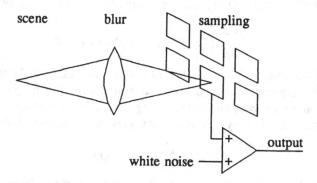

Figure 5.4: Simple camera model showing blur, sampling and noise

output can thus be treated as a virtual sensor which can then be incorporated into a Bayesian estimation framework. This approach permits us to take into account the spatially varying uncertainties which are often inherent in low-level visual processes. We expect that similar analysis can be applied to other low-level vision algorithms with similar benefits.

5.4 Dense data: image intensities

The sensor model for optical flow presented in the previous section, when combined with a dynamic Bayesian estimation framework, can provide surprisingly good depth estimates from small-motion sequences (Matthies *et al.* 1987). This approach, however, has some problems due mainly to the simplifying assumptions used in the flow estimator (these problems are discussed in Section 7.2). An alternative to using a separate flow estimation stage is to jointly estimate the intensity and flow fields (see Section 7.3). In order to perform this joint estimation, we need to develop a sensor model for the CCD camera used as our input device.

The model which we develop in this section, shown in Figure 5.4, has three components: blur, sampling, and noise. This model is thus quite simple, and does not include many of the known optical (geometric) distortions in the imaging process (Gremban *et al.* 1988). The blur induced by the optical system can be modeled by a point spread function $g(x, y)$. The shape of this function depends on the lens aperture, the focusing distance, and the distance to the object. The pattern of light falling on the CCD array can thus be computed as

$$f(x, y) = I(x, y) * g(x, y),$$

where $I(x, y)$ is the intensity array determined through the ideal perspective projection operation, and $*$ denotes convolution.

The output of one CCD cell can be computed by integrating over the area of the cell

$$o_{i,j} = \int_{-h/2}^{h/2} \int_{-w/2}^{w/2} f(i\Delta x + \lambda, j\Delta y + \eta) \, d\lambda \, d\eta,$$

where w and h are the width and height of the cell, and Δx and Δy are the array spacing. This integration is equivalent to first convolving $f(x, y)$ with a box filter $b(x, y)$, and then sampling the resulting signal. We can model the blur and sampling system together by convolving $I(x, y)$ with a new point-spread function

$$g'(x, y) = b(x, y) * g(x, y)$$

and then sampling the resulting signal. The advantage of this latter approach is that we can compute the Fourier transform of $g'(x, y)$, and thereby determine the aliasing inherent in the system.

For a system which performs a regular point-wise (Dirac delta) sampling of a filtered image, the amount of aliasing can be determined by examining the magnitude of the effective filter response in the areas outside the Nyquist frequency $(f_x, f_y) = (1/\Delta x, 1/\Delta y)$ compared to the magnitude inside (Rosenfeld and Kak 1976). Increasing the blur in the optics or using a larger cell size $w \times h$ will both reduce the aliasing, but at the expense of resolution. In practice, aliasing can affect computer vision algorithms such as depth-from-motion in several ways. Fine textures may appear to drift in the wrong direction if aliasing is present, and intensity edges appear to jump instead of slowly moving across the image. As we will see in Section 7.3, aliasing also makes precise (sub-pixel) positioning of edges difficult.

To overcome aliasing, we can de-focus the camera, but this may compound the problem of the variation of blur with depth. The best solution would be to design CCD sensors with larger sampling areas, or with a uniform blur close to the sensor surface. Another possible solution is to introduce *micro-jitter* in the imaging system. By rotating the camera around its optical axis over the range of one pixel, we can significantly reduce aliasing while only marginally reducing the resolution.

The final component of our simple camera model is the sensor noise. This noise can come both from the electrical noise present in the CCD array and from the quantization and distortion effects present in the digitization hardware. Using a white noise process results in the simplest possible model and estimation equations. More complicated noise models, such as those used by Geman and Geman (1984), could also be used within the Bayesian estimation framework.

To determine the validity of this simple model, we performed some experiments with a Sony XC-37 CCD camera with 16mm lens in the Calibrated Imaging Lab at Carnegie Mellon (Szeliski 1988b). We were able to compute the horizontal and vertical line spread functions by taking images of horizontal and vertical stripes which were slightly tilted with respect to the raster. The profiles obtained by averaging many aligned scan lines were then differentiated to obtain the line spread functions. Determining the aliasing effects was more difficult. The location of the midpoint of the step profile at each successive line did show some oscillation, but it was not sufficiently periodic to develop an aliasing model. This difficulty is probably due to the interaction between the CCD grid spacing and the digitizer sampling rate, which are different.

An additional effect that showed up in evaluating the camera model was a random horizontal shift of the pixels on successive scan lines. This jitter had previously been observed by Matthies *et al.* (1987) while developing an incremental depth-from-motion algorithm. To quantify the magnitude of this effect, a reference image was first obtained by averaging many images of the same static scene. A new image of the same scene was then taken, and each scan line was aligned to sub-pixel precision with the reference image using a least squares fit. The average magnitude of the observed jitter was about 0.10 pixels. While this jitter is not extreme, it is sufficiently large to affect the precision of the (Matthies *et al.* 1987) depth-from-motion algorithm, where the average inter-frame displacement is less than a pixel.

To reduce the amount of jitter, we can average several images of the same scene. A better solution is to use a CCD camera with a direct digital readout of each cell (traditional cameras convert the CCD array values to an NTSC analog signal). In addition to eliminating the jitter problem, using such a camera facilitates the calibration of the imaging sensor (Shafer 1988). The transfer function, blur function and aliasing of each cell can be measured separately. An alternative to this calibration procedure would be to enhance the CCD sensor with on-board circuitry that could "learn" to compensate for optical and geometric distortions by first being trained on a set of calibration data[4].

The development of a sensor model for image acquisition hardware thus serves two important functions. First, problems that are detected with the camera can often be compensated for by using pre-processing. Second, problems or uncertainties that cannot be eliminated can be modeled, and these models can be used by later stages of processing. In general, probabilistic modeling of sensors and low-level vision algorithms can greatly increase the accuracy of later processing stages and facilitate the optimal integration of information. Several examples of the benefits of this approach will be explored in the next chapter.

---

[4]This procedure could be analogous to the learning that occurs in the early visual pathways during childhood development. On-chip visual computation is rapidly becoming feasible (Sivilotti *et al.* 1987, Mead and Mahowald 1988).

Chapter 6

Posterior estimates

In the previous two chapters, we have developed a prior model for visible surfaces and a variety of sensor models for the inputs to low-level vision algorithms. In this chapter, we will see how the prior and sensor models can be combined using Bayes' Rule to obtain a posterior model. We will study how to compute optimal estimates of the visible surface from the posterior distribution. We will also show to calculate from this distribution the uncertainty inherent in a visible surface estimate, and discuss why such uncertainty modeling is important. Two novel algorithms which are based on the probabilistic posterior model will then be presented. The first algorithm estimates the regularization parameter λ from the sensed data using a maximum likelihood approach. The second algorithm estimates observer motion by matching sparse range data without using correspondence. These two algorithms illustrate the advantages of using a Bayesian approach to low-level vision.

6.1 MAP estimation

The probabilistic prior models and sensor models which we have studied in the previous two chapters are instances of Markov Random Fields. From the results which we obtained in Section 3.2, we know in this case that the posterior distribution is itself a MRF. This field can be described by a Gibbs distribution with an associated energy

$$E(\mathbf{u}) = E_p(\mathbf{u}) + E_d(\mathbf{u}, \mathbf{d}), \tag{6.1}$$

where $E_p(\mathbf{u})$ is the energy function associated with the prior model, and $E_d(\mathbf{u}, \mathbf{d})$ is the energy function that describes the sensor model. Computing the *Maximum A Posteriori* estimate is thus equivalent to minimizing the energy $E(\mathbf{u})$.

Several techniques can be used for performing this minimization, depending on the application. For surface interpolation or optical flow smoothing, the energy functions $E_p(\mathbf{u})$, $E_d(\mathbf{u}, \mathbf{d})$, and hence $E(\mathbf{u})$ are quadratic. Performing the minimization is thus equivalent to solving a large set of sparse linear equations. As discussed in Section 2.2, we can use one of several relaxation techniques to find the minimum energy solution. The advantage of using such iterative techniques over direct solution methods is that they can be implemented on massively parallel architectures with local connectivity. Terzopoulos (1984), Choi (1987), Blake and Zisserman (1987), and Szeliski (1989) present a number of different relaxation-based surface interpolation algorithms.

For stereo matching, the energy function being minimized has many local minima, so some search technique must be used. The most general (domain independent) approach is to use simulated annealing (Marroquin 1985, Szeliski 1986, Barnard 1989), which is a natural extension of the Gibbs Sampler algorithm discussed in Section 3.2. Alternative approaches include dynamic programming (Ohta and Kanade 1985) and scale space continuation (Witkin *et al.* 1987). Determining the location of discontinuities in the surface representation while performing surface interpolation also involves a non-convex energy minimization. A variety of continuation methods have been developed to perform this localization (Terzopoulos 1984, Koch *et al.* 1986, Blake and Zisserman 1987), and simulated annealing has also been used (Marroquin 1984).

The MAP estimate, however, is not the only estimate that can be computed from the posterior distribution $p(\mathbf{u}|\mathbf{d})$. As has been observed by Marroquin (1985), any loss function $L(\mathbf{u}, \mathbf{u}^*)$ can be used to define the optimal estimate. Given such a loss function, the optimal estimate $\mathbf{u}^*$ is the one that minimizes the expectation of loss

$$\langle L \rangle = \int L(\mathbf{u}, \mathbf{u}^*) p(\mathbf{u}|\mathbf{d}) \, d\mathbf{u}.$$

For MAP estimation, the loss function is a negative delta (we "win" one for guessing the correct estimate, but all other estimates are equally bad). A more sensible loss function for the terrain classification problem that we examined in Chapter 3 would be one which counts the number of misclassified pixels. This leads to the Maximizer of Posterior Marginals (MPM) estimator (Marroquin 1985)[1]. For many applications, we can also choose to compute the Minimum Mean Squared Error (MMSE) estimate[2].

The advantage of using loss functions to define the optimal estimate is that we can tailor the loss function to our particular application. In addition to allowing the development of task specific algorithms, this approach allows some top-down influence to be exerted on the low-level process. For example, in a robot navigation application where we want to avoid hitting obstacles, we can use a loss function that penalizes overestimates in distance more than underestimates.

---

[1] The MPM algorithm was used to generate the result shown in Figure 3.3d.

[2] For quadratic energy functions, the MAP and MMSE estimates are identical.

Using different loss functions can increase the power of probabilistic methods over simple energy minimization approaches. Having a single optimal estimate, however, still does not tell us how certain, accurate, or typical such an estimate might be. Ideally, we would like to pass the whole probability distribution on to the next level of processing. The Boltzmann Machine (Hinton and Sejnowski 1983) is an example of a system that provides such a distribution. In practice, however, we often have to restrict ourselves to a more parsimonious description. This forms the subject of our next section.

6.2 Uncertainty estimation

To characterize the uncertainty inherent in the output of a low-level vision algorithm, we can compute the second order statistics (covariance) of the estimate. These uncertainty estimates can be used to integrate new data, to guide search (set disparity limits in stereo matching), or to indicate where more sensing is required. For many distributions, second order statistics do not capture all of the useful information present in the distribution, but they are a good start. In this section, we will examine how uncertainty can be derived from the energy function characterizing the posterior distribution, and present two new algorithms for computing this uncertainty.

When we combine the regularization-based prior models developed in Section 4.1 with the simple sensor models developed in Section 5.1, we obtain posterior models that are Markov Random Fields with quadratic energy functions. This energy can be re-written in the form given in (2.16), i.e.,

$$E(\mathbf{u}) = \frac{1}{2}(\mathbf{u} - \mathbf{u}^*)^T A (\mathbf{u} - \mathbf{u}^*) + k, \qquad (6.2)$$

where $\mathbf{u}^*$ is the minimum energy solution. The Gibbs distribution corresponding to this quadratic form is a multivariate Gaussian with mean $\mathbf{u}^*$ and covariance A^{-1}. Thus, to obtain the covariance matrix, we need only invert the A matrix. One way of doing this is to use the multigrid algorithm presented in Section 2.2 to calculate the covariance matrix one row at a time. To obtain a single covariance field, we set $\mathbf{b} = \mathbf{e}_i$ in (2.14), i.e., we set all but one of the data points to 0 without modifying A, and solve as before.

Figures 6.1a and 6.1b show two covariance fields, one for the point in the lower right corner, and one for the point in the center. These fields are identical in shape (but not in magnitude) to the Green's functions (blending functions or shift-*variant* filters) that can be used to solve the interpolation problem. Their shape does not depend on the data values d_i, but only on the smoothing function (prior model) and the data confidence measures c_i. Intuitively, these covariance fields show how the overall surface would wiggle if one particular point was

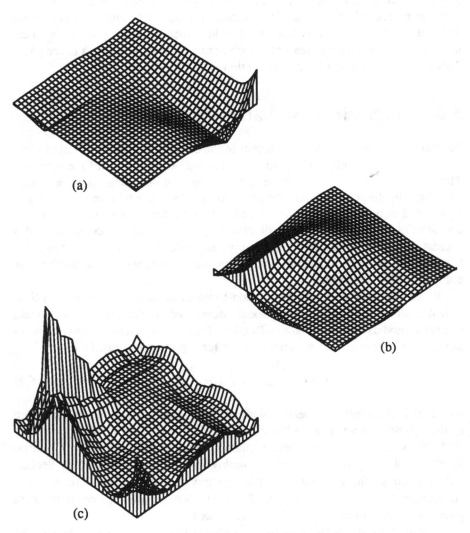

Figure 6.1: Sample covariance and variance fields

(a) and (b) sample covariance fields (c) variance field

moved up and down. For the special case of isotropic (shift-invariant) smoothing, the Green's function is equivalent to the smoothing filter $h_s(\mathbf{x})$ that is derived in Appendix B.

Storing all of the covariance fields is impractical because of their large size (for a 512×512 image, the covariance matrix has over 10^{10} entries). We can, however, keep only the variance at each point, i.e., the diagonal elements of the covariance matrix. These variance values are an estimate of the confidence associated with each point in the regularized solution (e.g., the residual uncertainty in optical flow *after* smoothing). Alternatively, they can be viewed as the amount of fluctuation of a point in the Markov Random Field (in the ensemble of typical fractal solutions). Figure 6.1c shows the variance estimate corresponding to the regularized solution of Figure 2.2c (this variance has been magnified for easier interpretation). The variance of the field increases near the edges and discontinuities; this is as expected, if we interpret the variance as the wobble or inverse global stiffness of the thin plate. We can thus develop a *dense* error model for the visible surface representation. Error modeling has not previously been applied to systems with such a large number of parameters.

Calculating the variance field using the above deterministic algorithm requires re-solving the system for each point in the field and is thus very time consuming. An alternative to this is to run the multigrid Gibbs Sampler at a non-zero temperature, and to estimate the desired statistics using a Monte Carlo approach. For example, we can estimate the variance at each point (the diagonal of the covariance matrix) by simply keeping a running total of the depth values and their squares. Unfortunately, the straightforward application of the Gibbs Sampler results in estimates that are biased or take extremely long to converge. This is because the Gibbs Sampler is a multidimensional version of the Markov random walk, so that successive samples are highly correlated, and time averages are ergodic only over very long time scales (even if the system is already at equilibrium). To help decorrelate the signal, we can use successive coarse-to-fine iterations and only gather a few statistics at the fine level each time. Examples of the variance field estimates obtained with such a stochastic algorithm are shown in Figure 6.2.

The stochastic estimation technique can also be used with systems that have non-quadratic (and non-convex) energy functions. In this case, the mean and covariance are not sufficient to completely characterize the distribution, but they can still be estimated. For stereo matching, once the best match has been found (say by using simulated annealing), it may still be useful to estimate the variance in the depth values. Alternatively, stochastic estimation may be used to provide a whole distribution of possible solutions, perhaps to be disambiguated by a higher level process.

Once we have calculated the variance field, we can use it to grow a confidence region around the mean or minimum energy solution. This confidence

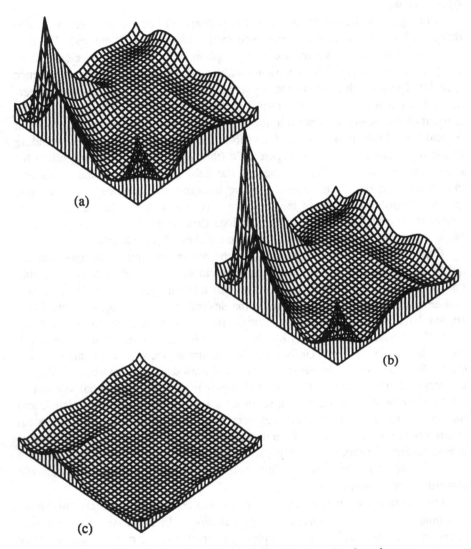

(a)

(b)

(c)

Figure 6.2: Stochastic estimates of covariance and variance

(a) multigrid, 1000 iterations (b) multigrid, 2500 iterations (c) single level, 1000 iterations

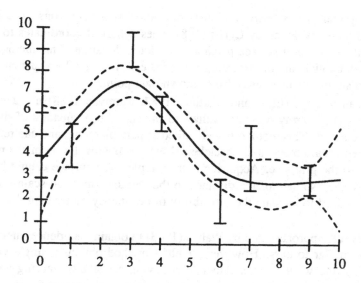

Figure 6.3: Cubic spline with confidence interval

region can be used in applications such as path planning or navigation. For example, we can use a 95% confidence interval if we wish to be "95% certain" of not hitting the surface. We can also look at the size of the confidence region (which is related to the local variance) to decide where additional active sensing may be required. Figure 6.3 shows a one-dimensional example of such a confidence region built around a cubic spline solution. Note that, just as for the thin plate, the uncertainty grows as we extrapolate away from known data.

The modeling of uncertainty in the visible surface representation is important if we are using this representation to aggregate data. In Section 6.4 we will examine the aggregation of sparse data taken from different viewpoints, while in Section 7.2 we will study the aggregation of dense depth measurements obtained from optical flow. Uncertainty modeling is also important if we will be matching the visible surface to models from an object database or to other intermediate representations. In Section 6.4 we will show how including such uncertainty modeling is essential to developing a surface-to-surface matching algorithm that can handle occlusions, limited areas of overlap, and sparse data. When matching to a model database, more accurate results may sometimes be obtained by directly matching the sensed points to the candidate model. However, it may often be easier to match the model to a dense intermediate representation, especially if a scale space approach (Witkin *et al.* 1987) is used to smooth the surface, and hence smooth the energy landscape of the matching algorithm.

The uncertainty representation scheme developed in this section differs in several important ways from previously developed representations. The spatial likelihood map developed by Christ (1987) uses spherical coordinates to represent the likelihood of a surface patch at a particular location. This method does not have an explicit smoothness constraint; instead, it uses a local planar model to extrapolate the surface away from known data points.

Occupancy maps (Elfes and Matthies 1987, Moravec 1988) use a two- or three-dimensional array of scalar values to indicate the occupancy of different portions of space. This method is well suited to path planning, and can represent three-dimensional objects and obstacles. However, the resolution of the method is limited to the grid size. Acquiring more samples of a surface cannot be used to improve the estimate of the distance to the surface, since the scalar value in each cell is used to indicate the probability of occupancy rather than the surface location.

Finally, the approach used by Wahba (1983) to obtain confidence intervals on splines is similar to ours. However, Wahba's method obtains a single variance estimate for the whole curve, rather than having a spatially varying variance. This method thus fails to capture some importance characteristics of the uncertainty, such as the increase in variance as we extrapolate away from known data.

In summary, we have shown how the second order statistics of the posterior estimate can be obtained by inverting the stiffness matrix $\mathbf{A}$. Two new algorithms have been presented to perform this inversion, the first being deterministic and sequential, the second being stochastic and parallel. We have also given some examples of how these uncertainty estimates can be used in later stages of visual processing. Additional examples will be given in this chapter and the next.

6.3   Regularization parameter estimation

One of the recurring problems associated with regularization and other energy-based estimation techniques is the need to select good values for the global parameters that control the algorithm. Some progress has been made in this direction (Craven and Wahba 1979), but mostly these parameters are still adjusted by hand. The advantage of using a Bayesian approach to low-level vision is that the unknown parameters can often be derived from knowledge of the problem domain or from the data itself. In Chapter 5, we saw how certain parameters can be computed from the noise characteristics of our sensors. In this section, we will examine how we can use the sensed data itself to estimate the remaining parameters.

Let us start by examining how certain global parameters affect the shape of the most likely and typical solutions of the surface interpolation problem. One possible parameterization for a regularization-based posterior estimate is

$$p(\mathbf{u}|\mathbf{d}) \propto \exp -[E_d(\mathbf{u}, \mathbf{d}) + \lambda E_p(\mathbf{u})]/T \qquad (6.3)$$

where λ controls the amount of smoothing, while T controls the roughness (amount of noise) in a typical sample from the MRF. Note that $1/T$ relates to the confidence ascribed to the data points, while λ/T encodes the variance in the prior model. Thus, if we have low confidence in the data, and wish the typical solution to be smooth, we use a high λ and a low T (Figure 6.4a); in the limit, this solution will approach a plane for the thin plate model. If we have a high confidence in the data, but still wish a typical solution to be smooth, we can use a low λ and a low T (Figure 6.4c); in the limit, this solution will approach an interpolating spline. Using higher T values does not affect the minimum variance solution, but typical surfaces have either more "crinkles" or more "wobble" (Figures 6.4b and 6.4d).

If the measurement noise can be determined from knowledge about the sensors, we can use the alternative parameterization

$$p(\mathbf{u}|\mathbf{d}) \propto \exp -[E_d(\mathbf{u}, \mathbf{d}) + E_p(\mathbf{u})/\sigma_p^2] \qquad (6.4)$$

where σ_p^2 simply encodes the overall variance in the prior model. While σ_p can be set on the basis of some prior knowledge about the application domain, it can also be determined from the data using *maximum likelihood* estimation. Intuitively, if σ_p is very low, then typical surfaces are extremely flat or planar, and it is unlikely that the given (non-flat) data sample would actually occur. Similarly, if typical surfaces are very rough, then the probability of a given data sample occurring becomes small. There exists some optimal value of σ_p that maximizes the probability $p(\mathbf{d})$ of actually having observed the given data.

To compute this probability, we will use the notation described in Appendix D which is derived from estimation theory. We start by modeling the prior as a multivariate Gaussian

$$\mathbf{u} \sim N(0, \mathbf{P}_0) \quad \text{with} \quad \mathbf{P}_0^{-1} = \sigma_p^{-2}\mathbf{A}_p$$

(σ_p^{-2} is equivalent to the regularization parameter λ). For our sensor model, we use the measurement matrix $\mathbf{H}$ to map from the surface $\mathbf{u}$ to the data points $\mathbf{d}$, and assume that the measurement noise is Gaussian

$$\mathbf{d} = \mathbf{H}\mathbf{u} + \mathbf{r}, \quad \text{with} \quad \mathbf{r} \sim N(0, \mathbf{R}).$$

The marginal distribution for $\mathbf{d}$ can be computed as

$$p(\mathbf{d}) = |2\pi(\mathbf{H}\mathbf{P}_0\mathbf{H}^T + \mathbf{R})|^{-1/2} \exp\left(-\frac{1}{2}\mathbf{d}^T(\mathbf{H}\mathbf{P}_0\mathbf{H}^T + \mathbf{R})^{-1}\mathbf{d}\right). \qquad (6.5)$$

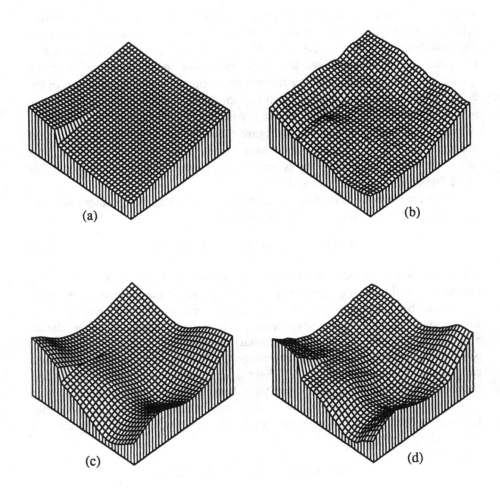

Figure 6.4: Typical solutions for various (λ, T) settings

(a) high λ low T (b) high λ high T (c) low λ low T (d) low λ high T

The probability of observing the data $p(\mathbf{d})$ is an implicit function of σ_p. To obtain the maximum likelihood estimate of σ_p, we simply maximize the value of $p(\mathbf{d})$ with respect to this parameter. Unfortunately, (6.5) is difficult to evaluate since it requires computing the whole covariance field of the prior model $\mathbf{P}_0$. Instead, we can use the simpler form derived in Appendix D, which allows us to write the negative logarithm of $p(\mathbf{d})$ as an energy function

$$E(\mathbf{d}) = -\log p(\mathbf{d}) = E_1(\mathbf{d}) + E_2(\mathbf{d}) \tag{6.6}$$

where $E_1(\mathbf{d})$ is the logarithm of the partition function and $E_2(\mathbf{d})$ is the energy (quadratic form) associated with the Gaussian. From (D.15), we have

$$E_1(\mathbf{d}) = \frac{1}{2}\log|\sigma_p^{-2}\mathbf{A}_p + \mathbf{H}^T\mathbf{R}^{-1}\mathbf{H}| - \frac{1}{2}\log|2\pi\mathbf{R}^{-1}| - \frac{1}{2}\log|\sigma_p^{-2}\mathbf{A}_p|, \tag{6.7}$$

and from (D.17) and (D.19), we have

$$E_2(\mathbf{d}) = \frac{1}{2}\mathbf{d}^T\mathbf{R}^{-1}(\mathbf{d} - \mathbf{H}\hat{\mathbf{u}}_1) \tag{6.8}$$

$$= \frac{1}{2}(\mathbf{d} - \mathbf{H}\hat{\mathbf{u}}_1)^T\mathbf{R}^{-1}(\mathbf{d} - \mathbf{H}\hat{\mathbf{u}}_1) + \frac{\sigma_p^{-2}}{2}\hat{\mathbf{u}}_1^T\mathbf{P}_0^{-1}\hat{\mathbf{u}}_1. \tag{6.9}$$

The log partition function $E_1(\mathbf{d})$ is minimized as $\sigma_p \to 0$ where its value approaches $\log|2\pi\mathbf{R}|$. Note that for a membrane or a thin plate, $\mathbf{A}_p$ is singular, so we must add a small diagonal element $\epsilon\mathbf{I}$ to this matrix when evaluating $E_1(\mathbf{d})$. The energy function $E_2(\mathbf{d})$, on the other hand, decreases as $\sigma_p \to \infty$, since in this case $\mathbf{u}^* \to \mathbf{d}$ as we approach the interpolated solution. Note that this energy can be written in two different ways (additional forms are shown in Appendix D). The first form, which is easier to compute, simply measures the weighted dot product of the data points and the residuals to the interpolated surface. The second form shows how this energy consists of a data compatibility term and a smoothness constraint, both of which are evaluated with respect to the minimum energy solution.

For a one-dimensional spline, the calculation of the matrix determinants is easy because the matrices are pentadiagonal (banded). This computation can be performed simultaneously with the solution of the pentadiagonal system. Consider the set of data points with known confidence intervals (standard deviations) shown in Figure 6.5. The cubic smoothing splines for various σ_p settings are shown in the same figure. For this example, we can easily evaluate the terms in (6.6) to obtain the log probability as a function of σ_p as shown in Figure 6.6. From these curves, we see that the maximum likelihood estimate of σ_p is about 5.

For two-dimensional interpolators, the calculation of the determinants is not as simple. We could use a direct algorithm to compute the determinant, but we

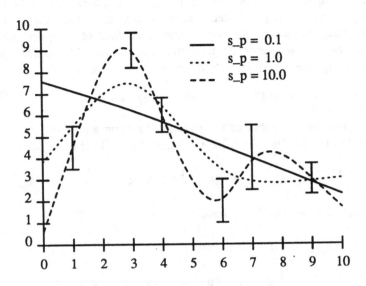

Figure 6.5: Family of splines of varying smoothness

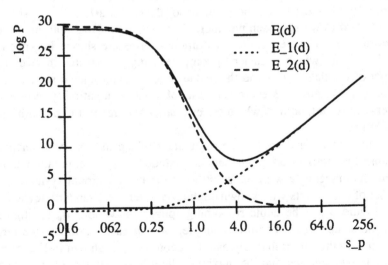

Figure 6.6: Maximum likelihood estimate of σ_p

would then have to keep intermediate results which are on the order of several image scan lines. An alternative is to analyze the asymptotic behavior of the matrix determinant as $\sigma_p \to 0$ and $\sigma_p \to \infty$, and to use these to approximate the true determinant.

When the data being smoothed is dense and uniform, we can express the probability distributions in the frequency domain. The sensed data **d** is then the sum of two Gaussian noise fields: the prior model, whose power spectrum is $\sigma_p^2|H_p(f)|^{-2}$, and the sensor noise, whose power spectrum is σ_d^2. The probability of the signal $d(x)$, whose Fourier transform is $D(f)$ is thus

$$p(D(f)) = \frac{1}{Z} \exp \left(-\frac{1}{2} \int \frac{|D(f)|^2}{\sigma_p^2|H_p(f)|^{-2} + \sigma_d^2} \right) df$$

with

$$\log Z = -\frac{1}{2} \int \log(\sigma_p^2|H_p(f)|^{-2} + \sigma_d^2)\, df.$$

We can use a maximum likelihood technique to estimate σ_p and σ_d, or we can just match the observed power spectrum $|D(f)|^2$ with the expected spectrum $\sigma_p^2|H_p(f)|^{-2}+\sigma_d^2$. The ability to compute the sensor noise and prior model variance simultaneously depends on the ability to separate out two noise processes with different spectral shapes (the prior model is *pink* or smooth, while the sensor noise is *white* or rough). This same method can also be used to estimate the fractal dimension of the surface in a manner similar to that proposed by Pentland (1984).

The maximum likelihood approach described in this section is just one possible method for estimating the desired degree of smoothing. The generalized cross-validation method developed by Craven and Wahba (1979) involves minimizing the distance between each data point d_i and the spline approximation fit to the remaining data points. For each sample value of λ, n spline fits must be calculated, where n is the number of data points; by contrast, our approach only requires a single fit. Spectral analysis has been used by Anderssen and Bloomfield (1974) to compute the optimal degree of smoothing. Their method estimates the prior model (signal) spectrum and noise spectrum from the spectrum of the data, and can thus only be used with regularly spaced measurements. Process whitening (Maybeck 1982, Marroquin 1984) chooses a smoothing model that makes the residuals between the data points and the surface as uncorrelated as possible. This approach, however, requires a large number of data points before its statistical assumptions can be justified. Geman and McClure (1987) use the expected value of the energy of the posterior estimate to calculate global parameter values. The exact theoretical connection between our new approach and these various methods and their relative performance remains to be investigated.

Two other intriguing possibilities associated with parameter estimation are learning and local smoothness estimation. Learning algorithms have recently had great success in teaching "neural nets" to associate input/output patterns (Ackley *et al*. 1985, Rumelhart *et al*. 1986) or even to perform simple visual tasks (Lehky and Sejnowski 1988). Since the amount of smoothing is characterized by a small number of global parameters, these should in theory be learnable; this would involve showing the surface interpolation system a sufficient number of examples of sparse input data and dense output solutions. The idea of locally estimating the degree of smoothness is suggested by the work of Pentland (1984), where the fractal dimension of the image is used to segment a scene. The resulting interpolator should behave similarly to the locally weighted bicubic splines studied by Foley (1987).

The Bayesian modeling approach to surface interpolation thus allows us to estimate the model parameters directly from the sampled data. The method which we have developed in this section estimates λ by finding the value that maximizes the likelihood of observing the given data $\mathbf{d}$. The negative logarithm of this likelihood, which we treat as an energy function, measures both the overall uncertainty in the model, which is encoded in the log determinants, and the fit between the data and the surface. The minimum of this energy function corresponds to the maximum likelihood estimate of λ. In principle, this same approach can be extended to estimate other model parameters, such as fractal dimension.

6.4 Motion estimation without correspondence

The probabilistic framework developed previously shows how a sparse set of measurements can be converted into a dense iconic map, and how the uncertainty in this map can be modeled and estimated. This same framework can be used to solve an extended version of the motion estimation problem: given two sets of points that come from the same surface but from different viewing directions, what is the most likely coordinate transformation between the two sets? This question is important in robot navigation and manipulation applications where the motion of the observer or object is to be determined.

Traditionally, motion estimation and pose determination problems have assumed that a correspondence is given or computable between the two sets of points to be matched (Ullman 1979, Webb and Aggarwal 1981). The problem is then to find a transformation $T(\mathbf{p}, \Theta)$ such that the distance between the transformed points and the original points is minimized (Tsai and Huang 1984, Faugeras and Hebert 1987). The new method presented in this section shows how to estimate this transformation even when no such correspondence exists. The two point sets can have a different number of points and limited areas of overlap. The approach is thus well suited for use with laser range finders or other active range sensors that do not sample the same points from different viewing position. It is also particularly well suited for terrain maps, since it can handle data points that are irregularly spaced (from perspective de-projection), and can incorporate prior knowledge from cartographic data. A more detailed description of the new algorithm is given in (Szeliski 1988a).

In describing our algorithm, we will use the same notation as in the previous section, except that we will subscript the data points d_k to indicate which set (or sensor position) they came from. Similarly, we will subscript the interpolated estimate $\hat{u}_k$ to indicate the best estimate after k data sets have been processed, and P_k to indicate its associated covariance[3]. The smoothness of the surface u is characterized by its prior covariance matrix P_0,

$$u \sim N(0, P_0),$$

and the data points are derived as some linear function of the surface corrupted by additive Gaussian noise

$$d_k = H_k u + r_k, \qquad r_k \sim N(0, R_k).$$

The measurement matrix H_k is used to convert the dense map u into a sparse measurement d_k. In general, the measurement vector d_k and the measurement matrix H_k depend (perhaps non-linearly) on some coordinate transformation parameter vector Θ_k. For now, we will assume that d_1 and H_1 are known, and that only the second set of measurements $d_2(\Theta)$ and measurement matrix $H_2(\Theta)$ are parameterized.

A simple approach for determining the motion parameters would be to interpolate the first set of data and to then measure the distance between the new set of points and the interpolated solution. We start by assuming that the first set of data points p_1 is registered with respect to the world coordinate frame. We then form an interpolated surface $u(x, y)$ by splitting each point into its location (x_i, y_i), which is incorporated into the H_1 matrix, and its depth value d_i, which becomes part of d_1. The variance assigned to each point is obtained by projecting the three-dimensional covariance matrix C_i onto the z-axis[4].

The second set of points p_2 is obtained by taking the sensor-based set of depth measurements p_2' and transforming them through a geometric transformation T which is parameterized by Θ

$$p_2 = T(p_2', \Theta).$$

These transformed points are then used to derive H_2, R_2 and d_2, which are now all functions of Θ. The exact form of Θ is not important here since we are not

---

[3]The subscript k corresponds to the discretized time variable used in Kalman filtering (Section 7.1).

[4]Another alternative is to use the force-field constraints developed in Section 5.2.

trying to obtain a closed-form solution. In this paper, we will use the translation vector (t_x, t_y, t_z) and three rotation angles $(\theta_x, \theta_y, \theta_z)$ as our parameters (see Tsai and Huang (1984) and Faugeras and Hebert (1987) for alternative formulations).

Assuming that the second set of data came from the interpolated surface $\hat{u}_1$, we can calculate the probability of the measurement vector d_2 being observed as

$$p(d_2 | \Theta) = |2\pi R_2|^{-1/2} \exp \left(-\frac{1}{2}(d_2 - H_2\hat{u}_1)^T R_2^{-1}(d_2 - H_2\hat{u}_1) \right). \qquad (6.10)$$

We can find the maximum likelihood solution for Θ by maximizing the above equation, or equivalently, minimizing the negative log likelihood

$$E_s(d_2) = -\log p(d_2|\Theta) = \frac{1}{2}\log|2\pi R_2| + \frac{1}{2}(d_2 - H_2\hat{u}_1)^T R_2^{-1}(d_2 - H_2\hat{u}_1) \quad (6.11)$$

(the subscript 's' stands for "simple"). In general, (6.11) does not have a closed-form solution for the minimum energy transformation (the vertical translation t_z is the exception). This energy equation must therefore be minimized using non-linear optimization techniques such as gradient descent (Press *et al.* 1986).

Unfortunately, using the above energy equation as the basis of our motion estimator has several problems. The matching of new data points to the extrapolated parts of the surface is inaccurate, since little is known about the surface in these areas. This is symptomatic of the more general problem with this technique, which is that it does not incorporate any knowledge about the uncertainty in the original interpolated surface. For example, range data will often have "shadowed" areas where the extrapolated data can be extremely uncertain. To overcome these problems, we must go back to the original Bayesian formulation and derive an optimal motion estimator.

To compute the optimal estimate of the motion, we find the value of Θ that makes it most likely that the two sets of data points p_1 and p_2 came from the same smooth surface. To compute the likelihood of observing the depth values d_2 (through the measurement matrix H_2), we note that the distribution $u_1 \sim N(\hat{u}_1, P_1)$ completely describes what is known about the smooth surface after the first set of data points has been incorporated. The second set of data points d_2 must thus be drawn from the distribution

$$d_2 \sim N(H_2\hat{u}_1, H_2P_1H_2^T + R_2)$$

(this same result can be obtained by writing down the joint probability function $p(u, d_1, d_2)$ and calculating the conditional probability $p(d_2|d_1)$). The negative log likelihood function in this case is

$$E_o(d_2) = \frac{1}{2}\log|2\pi(H_2P_1H_2^T + R_2)| + \frac{1}{2}(d_2 - H_2\hat{u}_1)^T(H_2P_1H_2^T + R_2)^{-1}(d_2 - H_2\hat{u}_1)$$
$$(6.12)$$

(the subscript 'o' stands for "optimal"). Unfortunately, (6.12) is difficult to evaluate since it involves the covariance matrix P_1, which is not sparse. However, using some algebraic manipulation (Appendix D), we can simplify the formula for the log likelihood to obtain

$$E_o(\mathbf{d}_2) = E_1(\mathbf{d}_2) + E_2(\mathbf{d}_2),\qquad(6.13)$$

where

$$E_1(\mathbf{d}_2) = \frac{1}{2}\log|2\pi\mathbf{R}_2^{-1}| + \frac{1}{2}\log|\mathbf{P}_1^{-1}| - \frac{1}{2}\log|\mathbf{P}_2^{-1}|\qquad(6.14)$$

and

$$E_2(\mathbf{d}_2) = \frac{1}{2}(\mathbf{d}_2 - \mathbf{H}_2\hat{\mathbf{u}}_1)^T\mathbf{R}_2^{-1}(\mathbf{d}_2 - \mathbf{H}_2\hat{\mathbf{u}}_2).\qquad(6.15)$$

The first component of the energy, E_1, measures the reduction in likelihood due to the sensor noise as traded off against the increase in posterior information. In practice, this component of the energy varies fairly slowly with the transformation parameter Θ and can usually be ignored. The second part of the energy, E_2, measures the distance between the new data points $\mathbf{d}_2$ and the surfaces $\hat{\mathbf{u}}_1$ and $\hat{\mathbf{u}}_2$. Note how (6.15) is similar to (6.11) except that one side of the quadratic form now involves the *new* surface estimate. Points that lie closer to the new surface estimate $\hat{\mathbf{u}}_2$ than to the old estimate $\hat{\mathbf{u}}_1$ are thus penalized less by the optimal energy measure. In this way, areas where the surface values are originally uncertain (because the data is uncertain, the area is shadowed, or the surface is being extrapolated) contribute less to the matching criterion.

A related energy measure which could be used instead of E_2 as a matching criterion is

$$E_3(\mathbf{d}_2) = \frac{1}{2}\hat{\mathbf{u}}_2^T\mathbf{P}_0^{-1}\hat{\mathbf{u}}_2 + \frac{1}{2}\sum_{i=1}^{2}(\mathbf{d}_i - \mathbf{H}_i\hat{\mathbf{u}}_2)^T\mathbf{R}_i^{-1}(\mathbf{d}_i - \mathbf{H}_i\hat{\mathbf{u}}_2)\qquad(6.16)$$

which is derived from the joint probability density function of the data and the surface (Appendix D). This new energy clearly shows the symmetry of the formula with respect to data set ordering. This alternate form also has a simple interpretation as the weighted distance between the data points and the optimal surface estimate (spring energy) plus the smoothness value of the surface (strain energy). The determination of an optimal motion estimate is thus equivalent to minimizing the energy of the composite surface and spring system.

The implementation of the optimal motion estimator is somewhat more complicated than simply matching the new data points $\mathbf{d}_2$ to the old surface estimate $\hat{\mathbf{u}}_1$. This is because the optimal surface estimate $\hat{\mathbf{u}}_2$ must be re-computed each time a new transformation Θ is generated. Fortunately, if successive transformations are close, the new surface estimate can be obtained from the previous estimate using only a few relaxation iterations. More importantly, since $\hat{\mathbf{u}}_2$ actually corresponds to the minimum energy solution of (6.16), we can jointly optimize $\hat{\mathbf{u}}_2$ and Θ using a continuation method similar to that used by Witkin *et al.* (1987).

The method which we have just described computes an optimal motion estimate by finding the transformation that minimizes an energy (negative log likelihood) function. This motion estimate is itself uncertain, i.e. it has a variance that can be determined from the shape of the error surface. We can justify estimating the distribution of the transformation Θ from the likelihood of $\mathbf{d}_2$ using Bayes' Rule,

$$p(\Theta|\mathbf{d}_2) = \frac{p(\mathbf{d}_2|\Theta)p(\Theta)}{p(\mathbf{d}_2)} \propto p(\mathbf{d}_2|\Theta)$$

if the prior distribution of the transformation $p(\Theta)$ is uniform. We can thus estimate a complete distribution for $p(\Theta|\mathbf{d}_2)$ by simply normalizing $p(\mathbf{d}_2|\Theta)$, which is the negative exponential of our energy function $E_o(\mathbf{d}_2)$. This distribution can encode multiple hypotheses about Θ by having multiple humps.

In practice, estimating a complete distribution for Θ may be impractical. Instead, we can take the optimal motion estimate $\hat{\Theta}$ and augment it with a covariance matrix $\mathbf{C}_\Theta$. One way of calculating $\mathbf{C}_\Theta$ is to first sample the complete distribution $p(\Theta|\mathbf{d}_2)$ and to then compute its first and second moments. A simpler, though potentially less accurate method is to simply fit a multi-dimensional parabola (second order polynomial) to the energy surface (see Szeliski (1988a) for details). The variance estimate thus obtained can be used to integrate the motion estimate provided by our new algorithm with other motion estimates, such as those provided by dead reckoning or inertial navigation systems.

The scheme we have just developed for estimating the motion between two frames can easily be extended to estimate a whole sequence of motions. The formulas given in (6.14) and (6.15) can be converted into a recursive form by simply substituting k for the subscript 2 and $k-1$ for the subscript 1. After each motion $\hat{\Theta}_k$ is computed, the new surface estimate $\hat{\mathbf{u}}_k$ and its inverse covariance $\mathbf{P}_k^{-1}$ are updated (we could also use the recursive formulas given in (D.9) and (D.10)). The Bayesian scheme which we have developed can also be used to incorporate domain-specific prior knowledge, such as cartographic information. The easiest way to do this is to introduce a phantom range observation before the first real set of data.

The method presented in this section has been tested on a number of noisy synthetic range images. Details of these experiments are reported in (Szeliski 1988a). To illustrate the behavior of our algorithm, we will present one simple example in this section. Figure 6.7a shows a very simple "blocks world" scene which has been sampled from two different directions (Figures 6.7b and 6.7c). The interpolated solution from the first set (Figure 6.8a) is quite inaccurate in the shadowed area behind the block, and yet the motion estimate obtained from

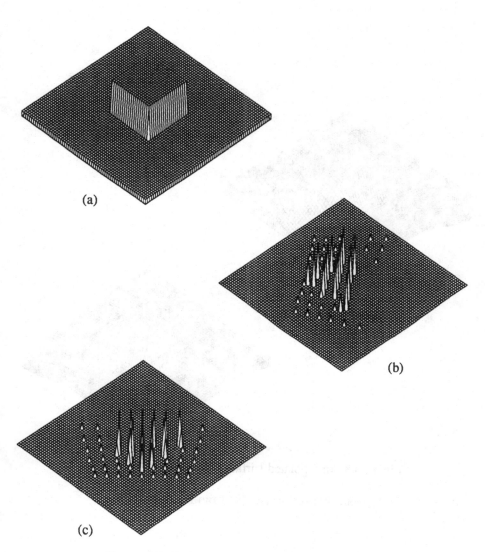

Figure 6.7: "Blocks world" synthetic range data
(a) synthetic scene (b) sparse range data from first viewing position (c) sparse range data from second viewing position

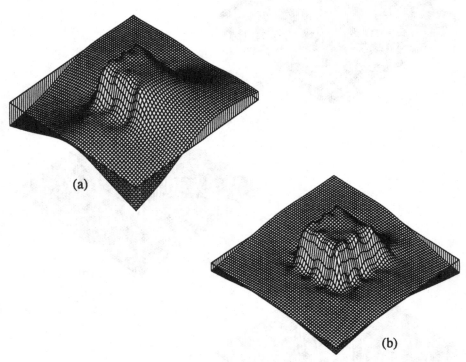

Figure 6.8: Interpolated surfaces from sparse block data

(a) from first set of data (b) from both sets of data

$$\Theta^* = \begin{bmatrix} 0.8000 \\ -0.1000 \\ -0.1000 \\ -0.0952 \\ -0.1000 \\ -1.6345 \end{bmatrix} \qquad \hat{\Theta}_o = \begin{bmatrix} 0.8005 \\ -0.1000 \\ -0.1000 \\ -0.0864 \\ -0.0975 \\ -1.6320 \end{bmatrix} \qquad \hat{\Theta}_s = \begin{bmatrix} 0.9100 \\ -0.1100 \\ -0.9980 \\ -0.1002 \\ -0.0950 \\ -1.6345 \end{bmatrix}$$

Figure 6.9: Motion estimate for blocks world data
The true motion is shown on the left, and the optimal estimate in the middle, and the simple method estimate on the right.

this data is still very good (Figure 6.9). The final surface reconstructed from the combined data is quite reasonable (Figure 6.8b).

We can compare this motion estimate to that obtained using the simple matching criterion given by (6.11). As shown in Figure 6.9, this estimate is quite far from the true motion. This is mostly due to the mismatch between the ground plane data seen from the second viewpoint (Figure 6.7c) and the solution interpolated from the first viewpoint in the shadowed area (Figure 6.8a).

The novel motion estimation algorithm presented in this section thus has several interesting features not present in previous algorithms. Our approach can handle sparse and irregularly spaced data and does not require any correspondence between sensed data points. The dense depth map that is built up incrementally becomes increasingly more accurate as more data is acquired. The depth map can also be extrapolated to shadowed or invisible areas without affecting the performance of the motion estimation algorithm, since our framework implicitly models the uncertainty in the interpolated surface. The uncertainty in the motion estimate can be calculated from the shape of the energy surface in the vicinity of the optimal estimate, and this information can then be used to integrate the motion estimate with other position sensors. The success of this algorithm in performing motion estimation further demonstrates the advantages of using a Bayesian approach to low-level vision.

Chapter 7

Incremental algorithms for depth-from-motion

The Bayesian models which we have developed in this book allow us to obtain optimal estimates of static visible surfaces, to integrate information from multiple viewpoints, and to analyze the uncertainty in our estimates. Many computer vision applications, however, deal with dynamic environments. This may involve tracking moving objects or updating the model of the environment as the observer moves around. Recent results by Aloimonos *et al.* (1987) suggest that taking an active role in vision (either through eye or observer movements) greatly simplifies the complexity of certain low-level vision problems. In this chapter, we will examine one such problem, namely the recovery of depth from motion sequences.

The study of depth-from-motion has long been an active area of research in computer vision. Early work concentrated on extracting the optical flow field from a pair of images, using either gradient-based (Horn and Schunck 1981) or correlation-based (Anandan 1984) techniques. More recent motion algorithms have attempted to use batch processing of the whole image sequence, either by fitting lines to the spatio-temporal data (Bolles and Baker 1985) or using spatio-temporal filtering (Adelson and Bergen 1985, Heeger 1986).

In this chapter we will develop an incremental algorithm which produces a dense on-line estimate of depth from the motion image sequence. The advantage of using the incremental approach is that coarse depth measurements are available immediately, and the quality of these estimates improves over time as more images are acquired. Incremental processing also has lower storage requirements than batch processing. The algorithm we develop uses an extended version of our Bayesian model which is based on the Kalman filter. In contrast to the algorithms which we have previously studied in this book, the depth-from-motion algorithm integrates the depth measurement with the surface interpolation stages, and thus provides a complete solution to a low-level vision problem.

In the first section of this chapter, we start by reviewing the Kalman filter approach to on-line estimation with dynamic systems. We then show how this framework can be applied to modeling visible surfaces, and discuss approximations needed to develop fast and simple algorithms. In the second section, we present an iconic depth-from-motion algorithm based on this framework which was developed by Matthies *et al.* (1987)(1989). We provide experimental results which show that the incremental approach to extracting depth from motion sequences yields good performance. We also discuss some problems with this algorithm, which are due mostly to the spatial and temporal correlations present in the area-based optical flow estimator. In the third section, we present a new incremental depth-from-motion algorithm which aims to overcome these problems. By jointly modeling the intensity and disparity fields, this algorithm can model the piecewise continuous nature of intensity images and obtain a performance which has the best features of both area- and edge-based depth-from-motion algorithms.

7.1   Kalman filtering

The Kalman filter is a Bayesian estimation technique used to track stochastic dynamic systems being observed with noisy sensors. The filter is based on three separate probabilistic models. The first model, the *prior model*, describes the knowledge about the system state $\hat{\mathbf{u}}_0$ and its covariance $\mathbf{P}_0$ before the first measurement is taken,

$$\mathbf{u} \sim N(\hat{\mathbf{u}}_0, \mathbf{P}_0). \tag{7.1}$$

As we have seen in Chapter 4, this model can capture the smoothness constraint associated with visible surfaces by setting $\mathbf{P}_0^{-1} = \mathbf{A}_p$. The second model, the *measurement* (or *sensor*) *model*, relates the measurement vector $\mathbf{d}_k$ to the current state through a measurement matrix $\mathbf{H}_k$ with the addition of Gaussian noise with a covariance $\mathbf{R}_k$,

$$\mathbf{d}_k = \mathbf{H}_k \mathbf{u}_k + \mathbf{r}_k, \qquad \mathbf{r}_k \sim N(0, \mathbf{R}_k). \tag{7.2}$$

When applied to surface estimation, the measurement matrix $\mathbf{H}_k$ is used to convert the dense map $\mathbf{u}_k$ into the sparse measurement $\mathbf{d}_k$. These two models form the basis for the Bayesian estimation framework which we have developed in this book. The Kalman filter introduces a third model, the *system model*, which describes the evolution over time of the current state vector $\mathbf{u}_k$. The transition between states is characterized by the known transition matrix $\mathbf{F}_k$ with the addition of Gaussian noise with a covariance $\mathbf{Q}_k$,

$$\mathbf{u}_k = \mathbf{F}_k \mathbf{u}_{k-1} + \mathbf{q}_k, \qquad \mathbf{q}_k \sim N(0, \mathbf{Q}_k). \tag{7.3}$$

In the case of depth-from-motion, the transition matrix F_k describes the mapping of surface estimates from one coordinate frame to the next as the observer changes position.

The above three models describe the evolution of the state u_k and its relationship to the measurements d_k. To obtain an optimal estimate $\hat{u}_k$ of the current state, the Kalman filter operates in two phases. The extrapolation phase predicts the new state given the previous best estimate

$$\hat{u}_k^- = F_{k-1}\hat{u}_{k-1}^+ \tag{7.4}$$

and updates the covariance matrix associated with the predicted estimate

$$P_k^- = F_{k-1}P_{k-1}^+ F_{k-1}^T + Q_{k-1}. \tag{7.5}$$

The correction phase updates the state estimate using the new measurements. First, we compute the Kalman filter gain matrix

$$K_k = P_k^- H_k^T [H_k P_k^- H_k^T + R_k]^{-1}. \tag{7.6}$$

We then update the state estimate by adding the residual between the observed and predicted measurements scaled by the Kalman filter gain

$$\hat{u}_k^+ = \hat{u}_k^- + K_k(d_k - H_k\hat{u}_k^-) \tag{7.7}$$

and reduce the covariance of the new estimate using

$$P_k^+ = (I - K_k H_k)P_k^-. \tag{7.8}$$

A block diagram for this implementation of the Kalman filter is shown in Figure 7.1.

Kalman filtering is usually applied to systems with a fairly small number of state variables. In the domain of motion sequence analysis, it has previously been used to track edges (Rives *et al.* 1986, Matthies and Shafer 1987, Baker and Bolles 1989, Baker 1989), but has not been used in conjunction with *dense* (iconic) fields such as depth maps. When modeling dense maps, the information matrices (inverse covariance matrices) are sparse and banded (because of the nature of the prior information matrix A_p), while the covariance matrices are not. A different formulation of the Kalman filter, which preserves the sparse nature of the matrices, is thus preferable.

Bierman (1977) discusses a number of efficient techniques for doing the Kalman filter update which rely on using matrix decomposition or factorization methods. These various decompositions (including the Square Root Information Filter) do not, however, result in matrices that are as sparse as the original information matrix. Thus for applications where the prior and posterior distributions are Markov Random Fields (of reasonable size), factorization methods are not

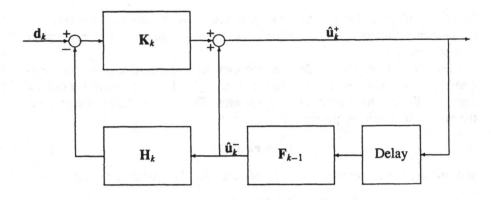

Figure 7.1: Kalman filter block diagram

useful since they require too much storage space ($O(n\sqrt{n})$ or $O(n^2)$) where n is the number of pixels).

For these reasons, we will use the the information matrix $A_k \equiv P_k^{-1}$ and the cumulative weighted data vector $b_k \equiv A_k\hat{u}_k$ as the quantities to be updated. The advantage of using this formulation is that the updating equations are particularly simple,

$$A_k^+ = A_k^- + H_k^T R_k^{-1} H_k^T \tag{7.9}$$

and

$$b_k^+ = b_k^- + H_k^T R_k^{-1} d_k, \tag{7.10}$$

and they require no matrix inversions (similar equations are used for sensor integration in Appendix D). The current estimate $\hat{u}_k^+$ can be computed at any time by solving

$$\hat{u}_k^+ = (A_k^+)^{-1} b_k^+$$

using multigrid relaxation. In practice, we can use the previous state estimate $\hat{u}_{k-1}^+$ or the predicted state estimate $\hat{u}_k^-$ as the starting point for the relaxation and only iterate for a few steps. This may not yield the optimal solution for the given data, but given enough time, the estimate will converge to such an optimal solution. Thus, a tradeoff can be made between the desired accuracy of the data and the amount of computation performed.

The prediction equations for our depth estimation system are somewhat more difficult to implement. This is because the mapping from one depth map u_{k-1} to the next is not a predetermined linear operation. Instead, the whole depth map is warped according to the local disparity to obtain the new depth map (Quam 1984). The exact form of this warping is explained in the next section. For now, let us assume that we can compute the transition matrix F_{k-1} by finding the linear mapping that defines how each point in the new field u_k is obtained as a weighted combination of points in the previous field $\mathbf{u}_{k-1}$ (the alternative is to use the *extended Kalman filter* (Gelb 1974) to compute $\mathbf{F}_{k-1}$ from the Jacobian of the state transition function).

Once we have computed the transition matrix $\mathbf{F}_{k-1}$, we can use (7.4) and (7.5) to calculate the predicted state $\hat{\mathbf{u}}_k^-$ and covariance $\mathbf{P}_k^-$. To keep the updating step simple, however, we would rather compute the predicted cumulative weighted data vector $\mathbf{b}_k^-$ and the information matrix $\mathbf{A}_k^-$. First, we will replace (7.5) with the simpler equation

$$\mathbf{P}_k^- = (1 + \epsilon)\mathbf{F}_{k-1}\mathbf{P}_{k-1}^+\mathbf{F}_{k-1}^T. \tag{7.11}$$

Thus, rather than assuming that the true system model is corrupted by additive Gaussian noise $\mathbf{q}_k$ (which seems strange for a static visible surface), we simply assume that our uncertainty in the old estimate (characterized by $\mathbf{P}_k$) uniformly increases by a small factor ϵ. In the Kalman filter literature, this is known as overweighting the most recent data (Maybeck 1982). Next, we re-write the information matrix prediction equation as

$$\begin{aligned}
\mathbf{A}_k^- &= (1 + \epsilon)^{-1}(\mathbf{F}_{k-1}(\mathbf{A}_{k-1}^+)^{-1}\mathbf{F}_{k-1}^T)^{-1} \\
&= (1 + \epsilon)^{-1}\mathbf{F}_{k-1}^{-T}\mathbf{A}_{k-1}^+\mathbf{F}_{k-1}^{-1}.
\end{aligned}$$

Finally, using (7.11) and (7.4) we write

$$\begin{aligned}
\mathbf{b}_k^- &= \mathbf{A}_k^-\hat{\mathbf{u}}_k^- \\
&= (1 + \epsilon)^{-1}(\mathbf{F}_{k-1}(\mathbf{A}_{k-1}^+)^{-1}\mathbf{F}_{k-1}^T)^{-1}\mathbf{F}_{k-1}(\mathbf{A}_{k-1}^+)^{-1}\mathbf{b}_{k-1}^+ \\
&= (1 + \epsilon)^{-1}\mathbf{F}_{k-1}^{-T}\mathbf{b}_{k-1}^+.
\end{aligned}$$

The above equations involve the inverse state transition matrix $\mathbf{F}_{k-1}^{-1}$. This inverse may in general be difficult to compute, although the sparse nature of $\mathbf{F}_{k-1}$ (since we are dealing with local map deformations) should make this easier. In particular, by assuming that the transformation is locally a translation, we can make the approximation

$$\mathbf{F}_{k-1}^{-1} \simeq \mathbf{F}_{k-1}^T,$$

since inverting the motion of a depth map is equivalent to interchanging the source and destination pixel locations.

One further simplifying assumption that we can make is that the prior model component of the information matrix $\mathbf{A}_k$ does not change as the observer changes viewpoint. This means that we keep the smoothness constraint expressed by $\mathbf{A}_p$ invariant, and only predict the certainties associated with the data points, which we can represent using

$$\tilde{\mathbf{A}}_k \equiv \mathbf{A}_k - \mathbf{A}_p.$$

Note that if the prior model were actually implementing a viewpoint-invariant smoother or interpolator (Blake and Zisserman 1986a), we could map this model through any rigid transformation without changing its behavior. The extension of these ideas to full three-dimensional models and general motion should be a promising area for further research.

The simplified Kalman filter for tracking visual surfaces can thus be summarized by the following prediction and updating equations:

$$\tilde{A}_k^- = (1 + \epsilon)^{-1} F_{k-1} \tilde{A}_{k-1}^+ F_{k-1}^T$$
$$b_k^- = (1 + \epsilon)^{-1} F_{k-1} b_{k-1}^+$$
$$\tilde{A}_k^+ = \tilde{A}_k^- + H_k^T R_k^{-1} H_k^T$$
$$b_k^+ = b_k^- + H_k^T R_k^{-1} d_k.$$

The current state estimate can be obtained by solving

$$\hat{u}_k^+ = (A_p + \tilde{A}_k^+)^{-1} b_k^+.$$

These simplified equations allow us to keep the amount of storage to a minimum, since b_k is a single field equal in size to the depth map u_k, and $\tilde{A}_k$ can be restricted to being diagonal (and is thus also a single field). The computations involved are simple, since the prediction step involves a warping of the data and inverse variance fields, and the updating involves a weighted addition. The need for matrix inversions is avoided, and the solution of the state estimate equation can be performed using multigrid relaxation.

The Kalman filtering framework which we have developed in this section is thus specially tailored to match the structure of the visible surface estimation problem. By using information matrices rather than covariance matrices, we can keep the representations sparse and the computations simple. However, since the state transition equations that describe the evolution of the retinotopic depth map are actually nonlinear, an optimal implementation would require the use of an extended Kalman filter (Gelb 1974). As we will see in the next section, however, we can still design a practical incremental depth-from-motion algorithm without resorting to this more complicated model.

7.2 Incremental iconic depth-from-motion

The Kalman filter-based framework described in the previous section has been used by Matthies, Szeliski and Kanade (1987, 1989) to develop an incremental image-based (iconic) depth-from-motion algorithm. In this section, we describe this algorithm, and show how it fits in with our general dynamic estimation framework. We then discuss the convergence properties of our estimator, and compare it to a feature-based (symbolic) algorithm which was also implemented. We present some quantitative performance results using a simple domain (a flat scene) and some qualitative results obtained with more realistic

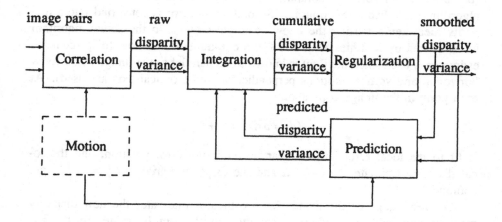

Figure 7.2: Iconic depth estimation block diagram

three-dimensional scenes. Lastly, we discuss some of the limitations of this approach, and preview how they may be resolved with a more sophisticated approach.

The algorithm developed by Matthies *et al.* (1987) consists of four main stages (Figure 7.2). The first stage uses correlation to compute an estimate of the displacement vector and its associated covariance. It converts this estimate into a disparity (inverse depth) measurement using the known camera motion. The second stage integrates this information with the disparity map predicted at the previous time step. The third stage uses regularization-based smoothing to reduce measurement noise and to fill in areas of unknown disparity. The last stage uses the known camera motion to predict the disparity field that will be seen in the next frame, and re-samples the field to keep it iconic (pixel-based).

The information propagated between these four stages consists of two fields (iconic maps). The first field is the disparity estimate computed at each pixel in the image. The second field is the variance associated with each disparity estimate. Modeling the variance at every pixel is essential because it can vary widely over the image, with low variance near edges and in textured areas, and high variance in areas of uniform intensity. These two fields roughly correspond to the cumulative data vector b_k and the point certainty matrix $\bar{A}_k$ used in the previous section. We will discuss how these representation differ, but first we will describe in more detail each of the four processing stages.

The first stage of the Kalman filter computes a disparity map from the difference in intensity between the current image and the previous image. In theory, this computation proceeds in two parts. First, a two-dimensional displacement (or optical flow) vector is computed at each point using the correlation-based algorithm described in Section 5.3. Second, this vector is converted into a disparity measurement using the known camera motion. In the actual algorithm implemented by Matthies *et al.* (1987), a one-dimensional flow (displacement) estimator is used since the flow is known to be parallel to the image raster (only horizontal and vertical motions perpendicular to the optical axis are used). At each point in the image, a parabola

$$e(d) = a\,d^2 + b\,d + c$$

is fit to the local error surface near its minimum error position, and the local disparity estimate $\hat{d} = -b/2a$ and the disparity variance $\sigma_d^2 = 2\sigma_n^2/a$ are computed.

The next stage of the iconic depth estimator integrates the new disparity measurements with the predicted disparity map (this step is omitted for the first pair of images). The algorithm assumes that each value in the measured or predicted disparity map is not correlated with its neighbors, so that the map updating can be done at each pixel independently. To update a pixel value, we first compute the variance of the updated disparity estimate

$$p_k^+ = \left((p_k^-)^{-1} + (\sigma_{d_k}^2)^{-1}\right)^{-1}$$

and then update the disparity value estimate with

$$u_k^+ = p_k^+ \left((p_k^-)^{-1}u_k^- + (\sigma_{d_k}^2)^{-1}\hat{d}_k\right).$$

This updating rule is simply a linear combination of the predicted and measured values inversely weighted by their respective variances.

The raw depth or disparity values obtained from optical flow measurements can be very noisy, especially in areas of uniform intensity. We use regularization-based smoothing of the disparity field in the third stage to reduce the noise and to fill in underconstrained areas. The smoothing algorithm we use is the generalized piecewise continuous spline under tension presented in Section 2.2. We use the inverse variance of the disparity estimate as the confidence associated with each measurement. The spline is computed using a three level coarse-to-fine multigrid relaxation method. This algorithm is amenable to implementation on a parallel computer, as are the other three processing stages. The variance field is not affected by the smoothing stage.

After the initial smoothing has been performed, depth discontinuities are detected by thresholding the angle between the view vector and the local surface normal (Matthies *et al.* 1987, Appendix B), and doing non-maximum suppression. Once discontinuities have been detected, they are incorporated into the piecewise continuous smoothing algorithm, and a few more relaxation steps are

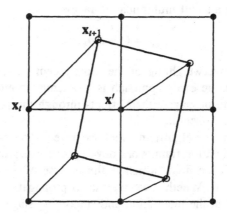

Figure 7.3: Illustration of disparity prediction stage

performed. Our approach to discontinuity detection, which interleaves smoothing and edge detection, is similar to the continuation method developed by Terzopoulos (1986b). The alternative of trying to estimate the boundaries in conjunction with the smoothing (Marroquin 1984, Leclerc 1989) has not been tried, but could easily be implemented within our framework. The detected discontinuities could also be propagated to the next frame, but this has not been implemented.

The final stage of the incremental depth-from-motion algorithm predicts the new disparity and variance fields from the previous fields and the observer motion estimate. At time k, the current disparity map and motion estimate are used to predict the optical flow between images k and $k + 1$, which in turn indicates where the pixels in frame k will move to in the next frame. In general this prediction process will yield estimates of disparity in between pixels in the new image (Figure 7.3), so we need to resample to obtain predicted disparity at pixel locations. For a given pixel x' in the new image, we find the square of extrapolated pixels that overlap x' and compute the disparity at x' by bilinear interpolation of the extrapolated disparities. Note that it may be possible to detect occlusions by determining where the extrapolated squares turn away from the camera. Detecting *disocclusions*, where newly visible areas become exposed, is not possible if the disparity field is assumed to be continuous, but is possible if disparity discontinuities have been detected.

The same warping and resampling algorithm used for the disparity is applied to the variance field. To partially model the increase in uncertainty which occurs in the prediction phase (due to uncertainty in the motion parameters, errors in calibration, and inaccurate models of the camera optics), we inflate the current variance estimates by a small multiplicative factor

$$p_{k+1}^- = (1 + \epsilon)p_k^+. \tag{7.12}$$

This corresponds to overweighting of the most recent data, as we saw in the previous section. A more exact approach is to attempt to model the individual sources of error and to propagate their effects through the prediction equations (Matthies *et al.* 1987, Appendix C).

The depth-from-motion algorithm which we have just described differs slightly from the dynamic estimation framework developed in the previous section. The algorithm uses a variance field, which is the inverse of $\bar{A}_k$ in the case where this matrix is diagonal. Whether we choose to propagate variance or inverse variance does not affect the integration and smoothing stages, but it does make a slight difference in the prediction stage because we must interpolate between the variances (or certainties). The algorithm also uses a disparity field instead of the weighted cumulative data vector $\mathbf{b}_k$. This field is actually more convenient, since it is easier to interpret, and can be used directly to predict the optical flow in the next frame. By feeding the result of the *smoothed* disparity field into the prediction stage instead of the *raw* field, however, the depth-from-motion algorithm actually oversmooths the data, since the measurements get re-smoothed every time they go around the prediction-integration-smoothing loop. This can be compensated for by decreasing the amount of smoothing used at each step and forgetting old data using (7.12). Despite the simplifications built into the depth-from-motion algorithm, the results obtained using this method are still very good, as we will see shortly.

7.2.1 Mathematical analysis

The depth estimates provided by the incremental depth-from-motion algorithm improve in accuracy as more images are acquired. Because the algorithm is based on a Bayesian formulation, and because we have a good model for the expected error in the flow estimates (Section 5.3), we can analytically derive the expected decrease in depth variance as a function of the number of frames. We can compare the convergence rate of our iconic algorithm to that of a symbolic (feature-based) algorithm and also to the results of stereo matching the first and last frames. Our results will be presented here using informal arguments. The full mathematical analysis can be found in (Matthies *et al.* 1987).

For the iconic method, we will ignore process noise in the system model and assume that the variance of successive flow measurements is constant. In this case, the disparity and variance updating equations we use in the depth-from-motion algorithm are equivalent to computing the average flow (Gelb 1974). From Appendix C, we know that the variance in an individual flow estimate $\hat{d}$ is

$2\sigma_n^2/a$, where σ_n is the standard deviation of the image noise and a is proportional to the average squared intensity gradient within the correlation window. If these flow measurements were independent, the resulting variance of the disparity estimate after t image pairs would be

$$\frac{2\sigma_n^2}{ta}. \tag{7.13}$$

However, the flow measurements are not actually independent. As we show in Appendix C, the error in the flow estimate $\hat{d}$ is proportional to the difference between the two error terms b_0 and b_1, which are proportional in turn to the average noise in the previous and current image. Generalizing this result from an image pair to an image sequence, we see that the correlation between two successive measurements is

$$\text{Cov}(\hat{d}_k, \hat{d}_{k+1}) = \left\langle (\hat{d}_k - \tilde{d})(\hat{d}_{k+1} - \tilde{d}) \right\rangle = \frac{1}{a^2} \left\langle (b_{k-1} - b_k)(b_k - b_{k+1}) \right\rangle = -\frac{\sigma_n^2}{a}.$$

With this correlation structure, averaging the flow measurements actually yields the following variance for the estimated flow:

$$\sigma_I^2(t) = \frac{2\sigma_n^2}{t^2 a}. \tag{7.14}$$

This result is interesting and rather surprising. Comparing equations (7.13) and (7.14), the correlation structure that exists in the measurements means that the algorithm converges faster than we first expected. An intuitive way of understanding this is shown in Figure 7.4. The disparity estimate obtained from each pair of images is equivalent to the slope of the dotted line segment connecting two position measurements. Of course, for the iconic method, we are not explicitly matching features; however, the error in each disparity (slope) estimate is still determined by the difference between two independent errors, one associated with each image. The average of all these slope measurements is thus the effective slope connecting the first and last positions. As we will see shortly, this is equivalent to the result obtained with stereo matching.

For the feature-based approach implemented by Matthies *et al.* (1987), both the sub-pixel position of each edgel and its disparity are jointly estimated by the Kalman filter. The resulting disparity estimate is equivalent to performing a least-squares fit to the edge positions as a function of time, as shown in Figure 7.4. The variance of the disparity estimate can be shown to be

$$\sigma_F^2(t) = \frac{12\sigma_e^2}{t(t+1)(t+2)} \tag{7.15}$$

where σ_e^2 is the variance in the edge position estimate. The variance of the displacement or flow estimate d thus decreases as the cube of the number of images.

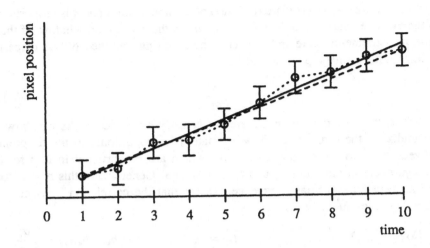

Figure 7.4: Computation of disparity using least squares fit

To compare these methods to stereo matching on the first and last frames of the image sequence, we must scale the stereo disparity and its uncertainty to match the flow between frames. This implies dividing the stereo disparity by t and the uncertainty by t^2. For the iconic method, we assume that the uncertainty in a stereo measurement will be the same as that for an individual flow measurement. Thus, the scaled uncertainty is

$$\sigma^2_{\text{IS}}(t) = \frac{2\sigma^2_n}{t^2 a}. \tag{7.16}$$

This is the same as is achieved with our incremental algorithm which processes all of the intermediate frames. For the feature-based approach, the uncertainty in stereo disparity is twice the uncertainty σ^2_e in the feature position; the scaled uncertainty is therefore

$$\sigma^2_{\text{FS}}(t) = \frac{2\sigma^2_e}{t^2}. \tag{7.17}$$

The results obtained with the incremental feature-based algorithm are thus more accurate than those obtained with stereo matching.

Extracting depth from a small-motion image sequence thus has several advantages over stereo matching between the first and last frames. The ease of matching is increased, reducing the number of correspondence errors. Occlusion is less of a problem, since it can be predicted from early measurements. Finally, better accuracy is available using the feature-based method. It would be nice if the iconic approach had the same improved accuracy. This is possible in theory, by extending our Kalman filter formulation to model the correlated nature of the

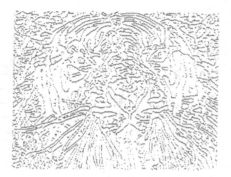

Figure 7.5: Tiger image and edges

measurements. We will examine a possible solution to this problem in Section 7.3.

7.2.2 Evaluation

The depth-from-motion algorithm described in this section has been tested on a number of image sequences acquired in the Calibrated Imaging Laboratory at Carnegie Mellon University. To measure the accuracy of our algorithm and to determine its rate of actual convergence, we needed a scene whose ground truth depth map was known. This was achieved by digitizing an image sequence of a flat-mounted poster[1]. Figure 7.5 shows the poster and the edges extracted from it. The ground truth value for the depth was determined by fitting a plane to the measured values, and the accuracy of the estimates was determined by computing the RMS deviation of the measurements from the plane fit. Since the scene used was flat, we could have reduced this RMS error to zero using a large degree of smoothing. For this reason, we did not use any regularization for our quantitative experiments. The results we present therefore show only the effect of the Kalman filtering algorithm.

To examine the convergence of the Kalman filter, the RMS depth error was computed after processing each image in the sequence for both the iconic algorithm described here and the feature-based algorithm described in (Matthies *et al.* 1987). We computed two sets of statistics, one for "sparse depth" and one for "dense depth." The sparse statistic is the RMS error for only those pixels where both algorithms gave depth estimates (i.e., where edges were found), whereas the dense statistic is the RMS error of the iconic algorithm over the full image. Figure 7.6 plots the relative RMS errors as a function of the number of

---

[1] Details of the actual physical setup (camera, lens, distance to scene and motion) are given in (Matthies *et al.* 1987).

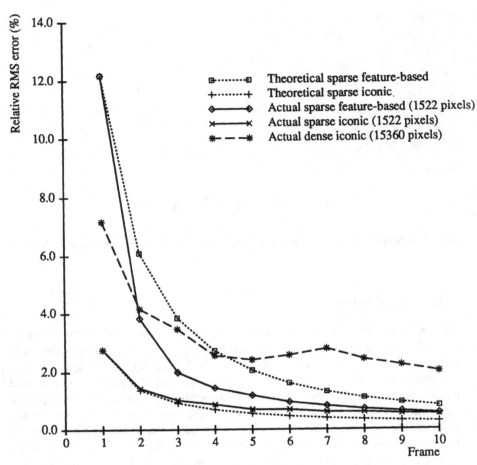

Figure 7.6: RMS error in depth estimate

images processed. Comparing the sparse error curves, the convergence rate of the iconic algorithm is slower than the feature-based algorithm, as expected. In this particular experiment, both methods converged to an error level of approximately 0.5% percent after processing eleven images. Since the poster was 20 inches from the camera, this equates to a depth error of 0.1 inches. Note that the overall baseline between the first and the eleventh image was only 0.44 inches.

To compare the theoretical convergence rates derived earlier to the experimental rates, the theoretical curves were scaled to coincide with the experimental error after processing the first two frames. These scaled curves are also shown in Figure 7.6. For the iconic method, the theoretical rate plotted is the quadratic convergence predicted by the correlated flow measurement model. The agreement between theory and practice is quite good for the first three frames.

Thereafter, the experimental RMS error decreases more slowly; this is probably due to the effects of unmodeled sources of noise. For the feature-based method, the experimental error initially decreases faster than predicted because the implementation required new edge matches to be consistent with the prior depth estimate. When this requirement was dropped, the results agreed very closely with the expected convergence rate. Finally, Figure 7.6 also compares the RMS error for the sparse and dense depth estimates from the iconic method. The dense flow field is considerably noisier than the flow estimates that coincide with edges, though there is just over two percent error by the end of eleven frames.

The iconic and edge-based algorithms were also tested on complicated, realistic scenes obtained from the Calibrated Imaging Laboratory. Two sequences of ten images were taken with camera motion of 0.05 inches between frames; one sequence moved the camera vertically, the other horizontally. The overall range of motion was therefore 0.5 inches; this compares with distances to objects in the scene of 20 to 40 inches. We will present some of the results obtained with the iconic algorithm here; a full discussion of the experimental results, including a comparison of the iconic and feature-based methods is given in (Matthies *et al.* 1987).

Figure 7.7 shows one of the images (a picture of a miniature town). Figure 7.8a shows a reduced version of the image and Figure 7.8b shows the intensity coded depth map produced by the iconic algorithm (lighter areas in the depth maps are nearer). This result was produced by combining disparity estimates from both the horizontal and the vertical camera motion sequences. The same depth map is shown in Figure 7.8c as a 3-D perspective reconstruction. Animated versions of the 3-D reconstructions proved to be very useful for detecting problems with our algorithms and for subjectively evaluating their performance. Figure 7.8d shows the occluding boundaries (depth discontinuities) found by the algorithm after the initial smoothing. The method found most of the prominent building outlines and the outline of the bridge in the upper left. From these figures, we can see that the main structures of the scene are recovered quite well.

Figures 7.9a and 7.9b show the results of our algorithms on a different model set up in the Calibrated Imaging Laboratory. The same camera and camera motion were used as before. Figure 7.9a shows the first frame, and Figure 7.9b shows the depth maps obtained with the iconic algorithm using both the vertical and horizontal motion sequences. Again, the algorithm did a good job in recovering the structure of the scene.

Finally, we present the results of using the first 10 frames of the image sequence used by Bolles *et al.* (1987). Figures 7.9c and 7.9d show the first frame from the sequence and the depth map obtained with the iconic algorithm. The results from using the feature-based method presented in (Matthies *et al.*

Figure 7.7: CIL image

1987) are similar to those obtained with the Epipolar-Plane Image technique. The iconic algorithm produces a denser estimate of depth than is available from either edge-based technique. These results show that the sparse (edge-based) batch processing algorithm for small motion sequences introduced by Bolles *et al.* (1987) can be extended to use dense depth maps and incremental processing.

The results obtained with the iconic incremental depth-from-motion algorithm are extremely encouraging. For the quantitative experiments performed with a flat scene, an accuracy of 0.5% was obtained after 11 images taken with a relatively narrow baseline[2]. The qualitative results obtained with different realistic toy model scenes are also suprisingly good. This algorithm is a powerful demonstration of the advantages of using a Bayesian modeling framework for solving low-level vision problems.

Nevertheless, there are some known problems and limitations with this al-

---

[2]This error figure "floor" may be due to the exponential weight decay that we use in updating the variances.

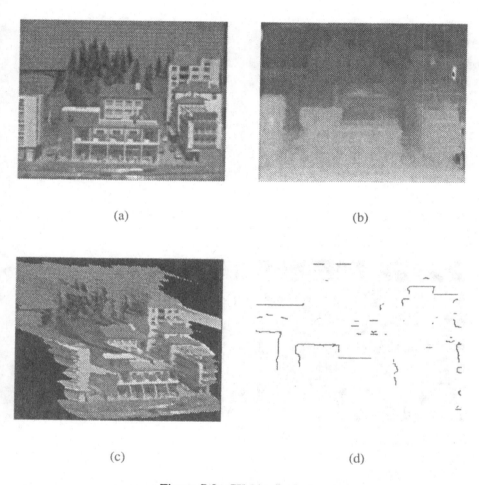

(a)   (b)

(c)   (d)

Figure 7.8: CIL depth maps
(a) first frame (b) combined motion depth map (c) perspective view (d) occluding
boundaries

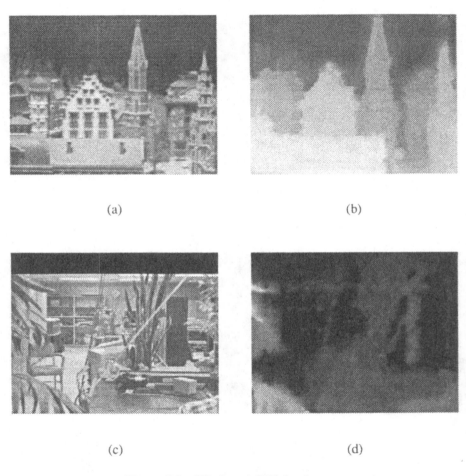

(a)   (b)

(c)   (d)

Figure 7.9: CIL-2 and SRI depth maps
(a) first frame from second CIL sequence (b) converged motion depth map (c)
first frame from SRI EPI sequence (d) horizontal motion depth map

gorithm. Because the temporal correlation between successive flow estimates is ignored, the optimal cubic convergence rate is not actually achieved. Because spatial correlations between measurements are ignored, the algorithm produces flat spots in the vicinity of strong image features such as intensity edges. The approximations used in developing the sensor model for the correlation-based flow estimator also tend to break down near such prominent features[3]. For these reasons, we present in the next section a new depth-from-motion algorithm which aims to overcome these limitations.

7.3   Joint modeling of depth and intensity

In this book, we have presented a general Bayesian framework for formulating low-level vision problems, and we have demonstrated the advantages of this approach over conventional methods. We have shown how prior models can capture the smoothness inherent in visible surfaces, how sensor models can describe the uncertainty in our measurements, and how dynamic system models can integrate information over time. In this section, we develop a new depth-from-motion algorithm which combines all of these ideas. Instead of using a pre-processing stage to extract flow from image pairs, we directly use the sampled images in our Bayesian formulation. More importantly, in addition to estimating the disparity field incrementally (as we did in the previous section), we also incrementally estimate the true intensity field.

The joint estimation of intensity and disparity has not previously been studied, yet it has the potential for significantly improving the accuracy of depth-from-motion algorithms. The usual *regularized stereo* formulation (Poggio *et al.* 1985a) assumes that the second image is a corrupted version of the first image, and uses an energy function which depends only on the intensity difference between the two images

$$E(\mathbf{d}) = \int \{[\nabla^2 G * (L(x, y) - R(x + d(x, y), y))]^2 + \lambda(\nabla d)^2\} \, dx \, dy.$$

In this section, we will argue that explicitly modeling and estimating the intensity field is preferable, because we can use better models of smoothness (e.g., piecewise continuous models), and because we can improve our estimates over time. The aim of our new formulation is to better model the visual world by using explicit prior models rather than implicitly building these assumptions into our algorithms. We also hope to improve the convergence rate of the incremental depth-from-motion algorithm by having two states (intensity and disparity) at each pixel instead of just one.

---

[3] Since the Taylor series expansion used in Appendix C is a particularly poor approximation to a step edge.

We start our presentation by re-examining the two-frame image matching problem, ignoring for the moment discontinuities in the intensity or disparity fields. Based on our probabilistic models, we derive a new equation for regularized stereo, and we show how the uncertainty in our estimates can be computed from variations in this energy. Next, we show how these estimates can be combined with a new image to obtain updated estimates of the intensity and disparity fields and their associated uncertainties. Lastly, we show how discontinuities can be integrated into our framework, and discuss possible extensions of this work.

7.3.1   Regularized stereo

The general depth-from-motion problem consists of processing a sequence of images taken from a moving camera in order to extract a depth map of the scene. For the work described here, we assume that the camera is moving horizontally (perpendicular to the optic axis) at a known constant velocity (extensions to other motions are discussed by Bolles *et al.* (1987)). Under these conditions, the intensity and disparity fields as a function of time can be described by

$$f^t(x + t \cdot d(x,y), y) = f(x,y) \tag{7.18}$$

$$d^t(x + t \cdot d(x,y), y) = d(x,y) \tag{7.19}$$

where $f(x,y)$ and $d(x,y)$ are the intensity and disparity fields at time $t = 0$. The evolution of the intensity field over time is shown in Figure 7.10. This Epipolar Plane Image (Bolles *et al.* 1987) shows the intensity of one row of an image as a function of time. The slope of the lines seen in this image corresponds to the disparity at those points. Of course the above two equations do not account for occlusions, but we will ignore this for now.

To model the smoothness in the intensity and disparity fields, we use the energy-based smoothness constraints introduced in Section 2.2. The disparity field is modeled by a thin plate, since it has the same smoothness as the visible surface that we are estimating[4]; the intensity field is modeled by a membrane[5]. We will designate the two functionals that measure this smoothness by $E_f(f)$ and $E_d(d)$.

The sensor model we use for our CCD camera is the one developed in Section 5.4,

$$g_i^t = (f^t * b)(x_i, y_i) + n_i^t, \qquad n_i^t \sim N(0, \sigma^2), \tag{7.20}$$

---

[4]For the camera motion we are assuming, the disparity is inversely related to depth. See (Matthies *et al.* 1987) for a discussion of the relative merits of disparity vs. depth representations.

[5]An informal justification for this choice is that if intensity is related to depth through the gradient (as it is for Lambertian reflection), then its degree of continuity should be one less than that of the depth itself.

Figure 7.10: An Epipolar Plane Image

where $b(x, y)$ is the two-dimensional blur function that captures both the blur induced by the optics and the integration over the finite area of the CCD sensor cell.

Putting these three models together, we obtain a new equation for regularized stereo

$$E(f, d) = \lambda_f E_f(f) + \lambda_d E_d(d) \tag{7.21}$$
$$+ \frac{1}{2\sigma^2} \sum_i \left[(b * f^1)(x_i, y_i) - g_i^1 \right]^2$$
$$+ \frac{1}{2\sigma^2} \sum_i \left[(b * f^0)(x_i, y_i) - g_i^0 \right]^2$$

where

$$f^1(x, y) = f(x, y) \quad \text{and} \tag{7.22}$$
$$f^0(x - d(x, y), y) = f(x, y) \tag{7.23}$$

(here we are estimating the intensity and disparity fields at time $t = 1$). Comparing this new energy equation to the formulation developed by Poggio *et al.* (1985a), we see that our formula does not contain any explicit smoothing of the intensity fields. This smoothing is achieved implicitly by the functional $E_f(f)$. The two image intensity samples g_i^1 and g_i^0 are not differenced directly, but are instead subtracted from the estimated intensity function $f(x, y)$. By warping the intensity function before differencing with the samples, our equation correctly models the non-uniform sampling induced by disparity gradients. Our formulation also differs from that of Witkin *et al.* (1987), who use normalized

cross-correlation as their similarity functional. Their approach is less sensitive
to bias or scaling in the images; if this were a concern, we could add global
parameters to the sensor model to account for these distortions. Our formulation
is related to that of Hung *et al.* (1988), who estimate the parameters of a surface
patch rather than a full disparity map.

To find the minimum energy solution of (7.22), we can use a variety of
techniques. We can take the correlation-based flow measurements described in
Section 5.3 as a starting solution, and then do gradient descent on the energy
function. Alternatively, we can use scale space continuation similar to that
of Witkin *et al.* (1987) by initially using larger values of λ_f (oversmoothing the
intensity estimates). Finding the minimum energy solution of (7.22) corresponds
to finding the MAP estimate. We can also run the system at a finite temperature
to find the mean (MMSE) estimate. It is likely that these two estimates are fairly
close, but this remains to be verified.

Before we can actually implement the energy minimization on a computer,
we must first discretize the energy equations. As usual, we represent both $f(x, y)$
and $d(x, y)$ as discrete fields $\mathbf{f} = \{f_i\}$ and $\mathbf{d} = \{d_i\}$ on a rectangular grid (we
will make these fields coincident with the sampled image $\mathbf{g} = \{g_i\}$). We use
the same discrete representations for the prior energies $E_f(\mathbf{f})$ and $E_d(\mathbf{d})$ as we
developed in Section 2.2. For computing the data compatibility equations, we
use bilinear interpolation[6] to convert from the discrete fields to the continuous
functions used in (7.22). We therefore write

$$f(x, y) = \sum_j f_j \, l(x - x_j, y - y_j)$$

where $l(x, y)$ is the basis function for the interpolator. From this, we compute

$$(b * f)(x_i, y_i) = \sum_j f_j \, (b * l)(x_i - x_j, y_i - y_j) = \sum_j h_{ij} f_j$$

where the h_{ij} coefficients encode how the information in measurement g_i is
spread among the nodal variables f_j. We now write the first data compatibility
term in (7.22) as

$$E_1(\mathbf{f}) = \frac{1}{2\sigma^2} \sum_i \left[\sum_j h_{ij}^1 f_j - g_i^1 \right]^2 = \frac{1}{2\sigma^2} (\mathbf{H}_1 \mathbf{f} - \mathbf{g}_1)^T (\mathbf{H}_1 \mathbf{f} - \mathbf{g}_1).$$

To compute the estimated intensity field in the previous frame $f^0(x, y)$, we
use (7.23) and find that $f^0(x, y)$ is still piecewise bilinear, but with a spacing that
is no longer regular. The basis functions $l_j^0(x, y)$ now depend on the structure of

---

[6]Using a higher order interpolator for $d(x, y)$ may improve the accuracy of the discrete
approximation, but also complicates the subsequent calculations.

the local disparity field around (x_j, y_j), and so we obtain

$$(b * f)(x_i, y_i) = \sum_j f_j (b * f_j^0)(x_i - x_j, y_i - y_j) = \sum_j h_{ij}^0(\mathbf{d})f_j$$

and

$$E_0(\mathbf{f}) = \frac{1}{2\sigma^2} \sum_i \left[\sum_j h_{ij}^0(\mathbf{d})f_j - g_i^1 \right]^2 = \frac{1}{2\sigma^2}(\mathbf{H}_0\mathbf{f} - \mathbf{g}_0)^T(\mathbf{H}_0\mathbf{f} - \mathbf{g}_0)$$

(note that $\mathbf{H}_0$ is a function of the disparity field $\mathbf{d}$).

Having derived the discrete energy equations, we now compute the partial derivatives and derive a quadratic approximation to the energy in the vicinity of the optimal solution. As we explained in Section 5.2, we then use this approximate energy to define a multivariate Gaussian distribution which represents our current estimate. If we are only interested in computing the uncertainty in the current depth estimate $\hat{\mathbf{d}}$, we can use the approximation

$$E'(\mathbf{d}) = E_d(\hat{\mathbf{d}}) + \sum_j c_j(d_j - \hat{d}_j)^2$$

where c_j can be approximated by $[f'(x_j, y_j)]^2/2\sigma^2$. However, we will model the correlation between the intensity and disparity estimates in order to obtain a faster convergence of the incremental algorithm.

Taking partial derivatives of

$$E(\mathbf{f}, \mathbf{d}) = E_f(\mathbf{f}) + E_d(\mathbf{d}) + E_1(\mathbf{f}) + E_0(\mathbf{f}, \mathbf{d}) \tag{7.24}$$

we obtain

$$\frac{\partial E}{\partial \mathbf{f}} = \mathbf{A}_f\mathbf{f} + \frac{1}{\sigma^2}\mathbf{H}_1^T(\mathbf{H}_1\mathbf{f} - \mathbf{g}_1) + \frac{1}{\sigma^2}\mathbf{H}_0^T(\mathbf{H}_0\mathbf{f} - \mathbf{g}_1)$$

$$\frac{\partial E}{\partial \mathbf{d}} = \mathbf{A}_d\mathbf{d} + \frac{1}{\sigma^2}\mathbf{f}^T\left(\frac{\partial \mathbf{H}_0^T}{\partial \mathbf{d}}\right)(\mathbf{H}_0\mathbf{f} - \mathbf{g}_1)$$

$$\frac{\partial^2 E}{\partial \mathbf{f}^T \partial \mathbf{f}} = \mathbf{A}_f + \frac{1}{\sigma^2}\mathbf{H}_1^T\mathbf{H}_1 + \frac{1}{\sigma^2}\mathbf{H}_0^T\mathbf{H}_0$$

$$\frac{\partial^2 E}{\partial \mathbf{d}^T \partial \mathbf{f}} = \frac{1}{\sigma^2}\left(\frac{\partial \mathbf{H}_0^T}{\partial \mathbf{d}^T}\right)(\mathbf{H}_0\mathbf{f} - \mathbf{g}_1) + \frac{1}{\sigma^2}\mathbf{H}_0^T\left(\frac{\partial \mathbf{H}_0}{\partial \mathbf{d}^T}\right)\mathbf{f}$$

$$\frac{\partial^2 E}{\partial \mathbf{d}^T \partial \mathbf{d}} = \mathbf{A}_d + \frac{1}{\sigma^2}\mathbf{f}^T\left(\frac{\partial^2 \mathbf{H}_0^T}{\partial \mathbf{d}^T \partial \mathbf{d}}\right)(\mathbf{H}_0\mathbf{f} - \mathbf{g}_1) + \frac{1}{\sigma^2}\mathbf{f}^T\left(\frac{\partial \mathbf{H}_0^T}{\partial \mathbf{d}}\right)\left(\frac{\partial \mathbf{H}_0}{\partial \mathbf{d}^T}\right)\mathbf{f}.$$

These partial derivatives are somewhat less daunting then they look. We can derive fairly simple formulas for the partials of $\mathbf{H}_0(\mathbf{d})$ by making simplifying assumption about the blur function $b(x, y)$. An alternative to performing this analysis is to use numerical techniques to estimate the partial derivatives. By measuring the increase in energy as we perturb the solution away from its optimal value, we can estimate these partial derivatives (this is analogous to measuring the local stiffness of the system).

The partial derivatives can be used to model our current estimate as a multivariate Gaussian with an energy function

$$E'(\mathbf{f}, \mathbf{d}) = \frac{1}{2} \left[(\mathbf{f} - \hat{\mathbf{f}})^T \quad (\mathbf{d} - \hat{\mathbf{d}})^T \right] \left[\begin{array}{cc} \frac{\partial^2 E}{\partial \mathbf{f}^T \partial \mathbf{f}} & \frac{\partial^2 E}{\partial \mathbf{f}^T \partial \mathbf{d}} \\ \frac{\partial^2 E}{\partial \mathbf{d}^T \partial \mathbf{f}} & \frac{\partial^2 E}{\partial \mathbf{d}^T \partial \mathbf{d}} \end{array} \right] \left[\begin{array}{c} (\mathbf{f} - \hat{\mathbf{f}}) \\ (\mathbf{d} - \hat{\mathbf{d}}) \end{array} \right].$$

Of course, we want to keep this representation as sparse as possible in order to minimize the amount of storage and subsequent computation that are required. Other than the prior matrices $\mathbf{A}_f$ and $\mathbf{A}_d$, we may be able to keep only the diagonal components, thus reducing these information matrices to iconic fields. For improved accuracy, it may be desirable to keep a few off-diagonal terms (e.g., between pixels that are adjacent horizontally). The exact tradeoffs involved will have to be determined empirically when the algorithm is actually implemented.

The new model for stereo matching presented here has several advantages over previous formulations. Because we are using a Bayesian framework, we can obtain better models for the sensors, model the non-uniform sampling effects due to disparity gradients, and explicitly specify the expected smoothness of our intensity and disparity fields. Using Bayesian models also allows use to compute the certainty in our disparity estimates. The quality of these certainty estimates is better than that obtained from the flow estimator used in the previous section, since we directly analyze the local variation of energy (log likelihood) with disparity. Most importantly, by modeling the cross-correlation between the intensity and disparity field estimates, we can integrate this stereo measurement with new images to obtain an on-line estimate of depth.

7.3.2   Recursive motion estimation

One of the shortcomings of the iconic depth-from-motion algorithm described in Section 7.2 is that in only attains a quadratic rate of convergence. By comparison, the symbolic depth-from-motion algorithm also studied by Matthies *et al.* (1987) is able to attain a cubic rate of convergence by tracking both the edge disparities and their sub-pixel positions. The joint modeling of intensity and disparity which we examine in this section has the potential to equal the accuracy of the symbolic method. As an informal justification, we note that if the local intensity gradient is constant, then the error in a disparity estimate d_j is proportional to the error in the intensity estimate f_j, which in turn depends on the error in the sampled intensities g_i near node j. By refining our estimate of f_j incrementally, and modeling the correlation between this estimate and the disparity estimate d_j, we are performing the same kind of two-state correlated estimation which leads to the cubic convergence rate of the symbolic method.

Implementing the new iconic depth estimation algorithm, however, is more difficult than implementing the symbolic matching algorithm, since we must account for the spatial correlations induced by our prior smoothness models and by other sources such as the blur in the sensor model. As we explained in Section 7.1, we can remove the prior model components from the information matrix before we perform the prediction, in order to keep the matrix structure more sparse and to make the smoothness constraint constant over time. One problem with the straightforward prediction equations developed in Section 7.1 is that in order to obtain a new disparity estimate, we need to measure the displacement between the current image and the old intensity estimate. For this reason, we will first present an update equation which avoids predicting intensity and disparity fields.

We model the old intensity and disparity estimates by $\hat{f}$ and $\hat{d}$, and the information matrix by

$$\begin{bmatrix} A_{ff} & A_{fd} \\ A_{df} & A_{dd} \end{bmatrix} = \begin{bmatrix} A_f + \bar{A}_{ff} & A_{fd} \\ A_{df} & A_d + \bar{A}_{dd} \end{bmatrix}$$

where the second form shows the partitioning of the information matrix into a prior model component and a current data certainty component. The energy equation for the new estimates f and d after incorporating the new measurements g_t is

$$E(f, d) = \frac{1}{2\sigma^2}(H_t f - g_t)^T(H_t f - g_t)$$

$$+ \frac{1}{2}\left[(f_{t-1} - \hat{f})^T \quad (d_{t-1} - \hat{d})^T\right] \begin{bmatrix} A_f + \bar{A}_{ff} & A_{fd} \\ A_{df} & A_d + \bar{A}_{dd} \end{bmatrix} \begin{bmatrix} (f_{t-1} - \hat{f}) \\ (d_{t-1} - \hat{d}) \end{bmatrix}$$

where f_{t-1} and d_{t-1} are derived from f and d by applying the system model equations corresponding to (7.23). Since we wish to have the smoothness constraints involving A_f and A_d expressed in terms of the current estimates f and d, we re-write the above equation as

$$E(f, d) = f_{t-1}^T A_f f_{t-1} + d_{t-1}^T A_d d_{t-1} + \frac{1}{2\sigma^2}(H_t f_t - g_t)^T(H_t f_t - g_t)$$

$$+ \frac{1}{2}(f_{t-1} - \tilde{f})^T \bar{A}_{ff}(f_{t-1} - \tilde{f}) + \frac{1}{2}(d_{t-1} - \tilde{d})^T \bar{A}_{dd}(d_{t-1} - \tilde{d}) \quad (7.25)$$

$$+ (f_{t-1} - \tilde{f})^T \bar{A}_{fd}(d_{t-1} - \tilde{d}) + k$$

where

$$\tilde{f} = \bar{A}_{ff}^{-1}(A_f + \bar{A}_{ff})\hat{f} = \hat{f} + \bar{A}_{ff}^{-1}A_f\hat{f}$$

$$\tilde{d} = \bar{A}_{dd}^{-1}(A_d + \bar{A}_{dd})\hat{d} = \hat{d} + \bar{A}_{dd}^{-1}A_d\hat{d}$$

are the estimates for intensity and disparity when the smoothing has been discounted. By replacing f_{t-1} and d_{t-1} by f and d in the first two terms of (7.25), we obtain the desired energy equation. From this new energy equation, we can compute the optimal intensity and disparity estimates, and repeat the variational analysis that we presented for regularized stereo to obtain an error model.

An alternative to this approach is to actually predict the flow and disparity fields from the old estimate, and to then compute the corrections that must be added to incorporate the new measurement. This approach is closer in spirit to the extended Kalman filter which we discussed in Section 7.1. In performing the prediction, we would separate out the prior model components of the information matrix, and preserve the sparseness of the remaining matrices. Which of the two approaches presented here is preferable remains to be seen.

The new incremental iconic depth-from-motion algorithm which we have developed here has the potential to achieve near-optimal estimates of intensity and depth from the image sequence[7]. It will be interesting to compare the results obtained with our algorithm to those obtained by "regularizing the whole depth-from-motion problem,"[8] i.e., obtaining a batch estimate for the dense intensity and disparity fields from the spatio-temporal block of data. It may be possible for scenes with no occlusion—such as the flat-mounted poster used in Section 7.2—to show that the results are equivalent.

7.3.3   Adding discontinuities

In the development of our new iconic depth-from-motion algorithm, we have so far omitted the discussion of intensity and disparity discontinuities. This omission was made to simplify the presentation. In fact, modeling the discontinuities in the two fields is one of the prime motivations for studying the joint modeling of intensity and disparity. The ability to use weak continuity constraints as prior models is a major advantage of the Bayesian estimation framework. These weak constraints better model the piecewise continuous nature of intensity images and visible surfaces.

To incorporate discontinuities into our prior models, we add line variables positioned on a dual grid (Section 2.4). We then modify the discretized energy equations to include these new variables, and add energy terms to the prior models $E_f(f)$ and $E_d(d)$ which penalize discontinuities (Appendix A). Previous algorithms which use discontinuity processes (Terzopoulos 1984, Marroquin 1984, Blake and Zisserman 1987) always represent these as binary variables; for our application, we will also represent the sub-pixel positions of the dis-

---

[7] A linear or linearized filter such as the Kalman filter cannot give optimal performance on a non-linear problem such as depth-from-motion.

[8] Suggested by Andrew Witkin.

continuities. Estimating the sub-pixel edge position will allow us to use the symbolic matching algorithm developed by Matthies *et al.* (1987) to obtain a cubic convergence rate for the depth estimates at these locations.

The sub-pixel position of the edge can be represented by attaching a single real value to each binary line variable. A more iconic representation for this position is the *interpolation coding* scheme developed by Ballard (1987), which makes the warping operations used during the prediction phase more uniform. The estimation of the sub-pixel position can be done in conjunction with the line variable estimation or as a post-processing stage. We can use the Graduated Non-Convexity (GNC) algorithm of Blake and Zisserman (1987) for the intensity fields, since it is particularly well suited to finding discontinuities in dense input data. We can compute the edge position and its uncertainty from the shape of the blur function, the camera noise, and the local intensity values around the edge.

The estimation of disparity discontinuities is more difficult, since they cannot be directly inferred from the intensity images, and may even be difficult to infer from the disparity map. This is especially true if the disparity map itself is sparse, which is often the case with stereo or motion-based depth estimates. One approach that may help is to link the likelihood of disparity edges to the existence of intensity edges (Gamble and Poggio 1987); this fits in well with our Bayesian modeling approach. Another possibility is to try to detect occlusions or disocclusions in the motion sequence. For each depth discontinuity that is found we can assign the depth estimate associated with that edge to the upper (nearer) surface bounded by the edge. Alternatively, we can use a separate depth value variable for each discontinuity that is detected.

Much work needs to be done to convert the ideas presented in this section into a working algorithm. The potential for improved accuracy of such an algorithm over existing depth-from-motion estimators, however, is promising. By including discontinuity processes into our intensity and disparity models, we obtain a more veridical description of the scene, and also improve the convergence rate of our depth estimates. The Bayesian formulation of depth-from-motion allows us to use better sensor models, to model the non-uniform sampling induced by disparity gradients, and compute the certainty associated with our estimates. We can also extend this framework to model other intrinsic images. For example, we could estimate the reflectance functions of the surfaces rather than the intensities, and thus account for the variation of intensity with viewer position. Using full three-dimensional models instead of retinotopic maps would also increase the descriptive power of our system, and allow us to "remember" the surfaces that become temporarily occluded during a motion sequence (Figure 7.10).

The dynamic Bayesian estimation framework which we have developed in this chapter is a good demonstration of the advantages of Bayesian modeling for low-level vision. Using the Kalman filter, we can accumulate measurements over time and improve the accuracy of our estimates. As we have shown, however, we must be careful about choosing our representations in order to produce a feasible algorithm. The iconic depth-from-motion algorithm which we reviewed in Section 7.2 is a practical demonstration of the utility of these ideas and of their applicability in real-world domains. The new algorithm which we have proposed in this section further exploits the power of Bayesian modeling, and suggests that significant improvements in descriptive power and accuracy are possible with this approach.

Chapter 8

Conclusions

In this book, we have developed a Bayesian model for the dense fields that arise in low-level vision, and shown how this model can be be applied to a number of low-level vision problems. We have used this model to analyze the assumptions inherent in existing vision algorithms, to improve the performance of these algorithms, and to devise novel algorithms for problems which have not previously been studied. In this chapter, we will summarize these important results and discuss how they can be extended in the future to other computer vision problems.

8.1 Summary

The main focus of this work has been the development of a Bayesian framework for modeling dense fields and their associated uncertainties. Such fields are used in low-level vision to represent visible surfaces and intrinsic images. These retinotopic maps form a useful intermediate representation for integrating information from different low-level vision modules and sensors. Modeling the uncertainty in these maps is an essential component of the integration process, and provides a richer description for later stages of processing.

The Bayesian framework we have developed is based on three separate probabilistic models. The prior model describes the *a priori* knowledge that we have about the structure of the visual world. The sensor model describes how individual measurements (such as image intensities) are obtained from a particular scene. The posterior model is derived from the first two models using Bayes' Rule and describes our current estimate of the scene given the measurements. By examining each of these models in turn, we have developed new algorithms for low-level vision problems as well as providing new insights into existing algorithms.

In studying the prior model, we have analyzed the statistical assumptions in regularization-based smoothing, developed a new graphics algorithm, and developed a new multiresolution representation. The prior model captures the smoothness or coherence assumptions associated with a visible surface. We construct this model by using the smoothness constraint (stabilizer) from regularization to define the energy function of a Markov Random Field. Using Fourier analysis, we show how the choice of the stabilizer used in regularization determines the power spectrum (and hence the correlation function) of the prior model. For the membrane and the thin plate—two of the most commonly used regularization models—we show that the resulting prior model describes a random fractal surface. This result leads us to a new understanding of the role of regularization: the choice of a particular stabilizer (degree of smoothness) is equivalent to assuming a particular power spectrum for the prior model.

Using the regularization-based probabilistic prior model, we have devised two new algorithms for modeling surfaces. The first algorithm uses multigrid stochastic relaxation to generate fractal surfaces for computer graphics applications. These surfaces can be arbitrarily constrained with depth and orientation constraints and discontinuities; our method thus exhibits a degree of flexibility not present in previous algorithms. Our second algorithm is a relative multiresolution representation for visible surfaces. This representation encodes local deviations in depth at each level of the pyramid, with the sum of all the levels defining the absolute depth map. We have shown how the power spectrum of the composite representation can be computed by summing the power spectra of the individual levels. This gives us a new technique for shaping the frequency response characteristics of each level and for ensuring the desired global smoothing behavior of the system.

In studying sensor models, we have developed a new constraint for sparse depth measurements and analyzed the uncertainty in optical flow and intensity images. Starting with the equivalence between a point sensor with Gaussian noise and a simple spring constraint, we show how to extend this model to other one-dimensional uncertainty distributions. We then develop a new sensor model which incorporates the full three-dimensional uncertainty associated with a sparse depth measurement. The constraint corresponding to this model acts like a force field or a slippery spring, and can thus be used in conjunction with parametrically defined surfaces. We show how to approximate this non-linear constraint by a quadratic energy function which fits into our visible surface representation.

We have also applied sensor modeling to systems which produce dense measurements. First, we analyze the uncertainty associated with correlation-based optical flow measurements. The uncertainty in the estimates is derived from the shape of the local error surface, thereby accounting for the spatially varying reliability of optical flow estimates. These flow measurements, along with their uncertainties, are used as part of an iconic depth-from-motion algorithm. Second, we develop a simple stochastic model for a CCD camera which describes the blur, sampling and noise inherent in the imaging system.

In studying the posterior model, we have shown how to calculate the uncertainty in the posterior estimate from the energy function of the system, and we have developed two new algorithms to perform this computation. The first algorithm uses deterministic relaxation to calculate the uncertainty at each point separately. The second algorithm generates typical random samples from the posterior distribution and calculates statistics based on these samples. The uncertainty map obtained from these algorithms can be used to set confidence limits on our measurements or to suggest where further active sensing is required.

Using the probabilistic description of the posterior estimate, we have developed two new parameter estimation algorithms. The first algorithm estimates the optimal amount of smoothing to be used with regularization. This estimate is obtained by maximizing the likelihood of the data points that were observed given a particular (parameterized) prior model. The second algorithm determines observer or object motion given two or more sets of sparse depth measurements. The algorithm determines this motion—without using any correspondence between the sensed points—by maximizing the likelihood that point sets came from the same smooth surface.

Finally, we extend our Bayesian model to temporal sequences using a two-dimensional generalization of the Kalman filter. By paying careful attention to computational issues and to alternative representations, we obtain simple formulations for the updating equations. Using this framework, we have developed two new depth-from-motion algorithms.

The first algorithm estimates optical flow from successive pairs of images and incrementally refines the resulting disparity estimates and their associated confidences. This algorithm produces a dense on-line estimate of depth which improves over time. Experiments with real images have demonstrated the improved accuracy which can be obtained with this approach, and shown that the reconstructed depth maps of the scene are quite realistic. The second depth-from-motion algorithm jointly models the intensity and disparity fields along with their discontinuities. By using explicit prior and sensor models, this algorithm avoids some of the problems with existing techniques, such as the correlation between measurements in flow-based methods and the difficulty in dealing with disparity gradients. These two algorithms demonstrate the advantages of applying Bayesian modeling to low-level vision problems.

8.2 Future research

The Bayesian framework we have developed has thus far only been applied to visible surfaces (2½-D sketches) and to surface interpolation and depth-from-motion. In future work, we plan to extend our Bayesian approach to other visual representations and to other computer vision problems. These include the extension to full three-dimensional models and to multiple intrinsic images as well as the development of more general depth-from-motion and shape-from-x algorithms.

The extension to viewpoint-invariant surface models and energy-based three-dimensional models is the most straightforward. To perform this extension, we use the smoothness energy associated with the surfaces to define the prior model through the Gibbs distribution. While the resulting distributions are no longer correlated Gaussians (because the energy functions are not quadratic), they are still Markov Random Fields (because of the local structure of the energy). Applying the Bayesian approach to these representations should produce similar benefits to those which we have demonstrated in this book, including the ability to use better sensor models and the ability to characterize the uncertainty in the estimates. We also plan to examine the extension to locally tensioned splines and to non-spline models such as the constant curvature sign models suggested by Blake and Zisserman (1987).

Three-dimensional elastic net models are an ideal application for the new depth constraint developed in Section 5.2. We plan to study how these constraints can be used to fit three-dimensional models either to sparse data, such as that available from direct range sensing or tactile sensing, or to contour information. The advantage of the elastic net, which is a parametric surface, coupled with the force field depth constraint, is that we need not establish a correspondence between data and surface points.

We also plan to apply the Bayesian modeling approach to multiple intrinsic images, thus providing a unified framework for describing many different low-level vision algorithms. For example, we can extend our new depth-from-motion algorithm by estimating the reflectance functions (albedos) of the visual surfaces, and thus incorporate shading cues into the reconstruction process. We would also like to study the more general idea of intrinsic models—probabilistic descriptions of intrinsic images—and how to link these models to higher level three-dimensional models. In particular, we should examine how to use the uncertainty in the intermediate level estimates to determine the uncertainty in the three-dimensional model parameters. For this approach to be viable, however, we will first have to solve the problems of grouping, segmentation, and discontinuity detection.

The relative multiresolution depth representation which we have introduced needs further refinement. In future work, we will develop efficient parallel algorithms for the solution of the interpolation problems using this representation, and study how to assign the discontinuities to the appropriate level. We will also examine how to integrate information from local depth cues such as disparity gradients and how to use multiresolution inputs derived from frequency channel (band-pass) descriptions of the image.

The extension of of our depth-from-motion algorithm to general motion is another area of future research. Combining this idea with full three-dimensional models, we could construct an active vision system which builds a three-dimensional description of its environment by roaming around. The Bayesian modeling of surfaces which we have developed in this book would be an essential component of such a system, allowing information from many viewpoints and sensor modalities to be integrated in a natural and statistically optimal fashion. This statistical framework could also be used to jointly estimate the scene description and the observer motion, which is a much more difficult problem.

The representations and algorithms which we have been studying are all examples of massively parallel computation. By implementing these algorithms on parallel computers, we will obtain significant speed-ups in operation and also shed light on desirable characteristics for new computer architectures. Ultimately, some of the low-level operations will be implemented directly in silicon using VLSI technology, perhaps even using analog computations.

In conclusion, we have developed in this book a Bayesian approach for estimating visual surfaces and other two-dimensional fields. The modeling of surfaces using this approach has several advantages. We have used Bayesian models to compute the uncertainties associated with visible surface estimates and to develop a number of novel robust low-level vision algorithms. By demonstrating that these algorithms can be used to solve real-world vision problems more efficiently than currently existing techniques, we have established that Bayesian modeling is a powerful and practical framework for low-level vision.

Bibliography

Ackley, D. H., Hinton, G. E., and Sejnowski, T. J. (1985). A learning algorithm for Boltzmann Machines. *Cognitive Science, 9,*, 147–169.

Adelson, E. H. and Bergen, J. R. (1985). Spatiotemporal energy models for the perception of motion. *Journal of the Optical Society of America, A 2(2)*, 284–299.

Adler, R. J. (1981). *The Geometry of Random Fields.* J. Wiley, Chichester, England.

Aloimonos, J., Weiss, I., and Bandyopadhyay, A. (1987). Active vision. In *First International Conference on Computer Vision (ICCV'87)*, pages 35–54, IEEE Computer Society Press, London, England.

Anandan, P. (1984). Computing dense displacement fields with confidence measures in scenes containing occlusion. In *Image Understanding Workshop*, pages 236–246, Science Applications International Corporation, New Orleans, Louisiana.

Anandan, P. and Weiss, R. (1985). Introducing a smoothness constraint in a matching approach for the computation of displacement fields. In *Image Understanding Workshop*, pages 186–196, Science Applications International Corporation, Miami Beach, Florida.

Anderssen, R. S. and Bloomfield, P. (1974). A time series approach to numerical differentiation. *Technometrics, 16(1)*, 69–75.

Anstis, S. M. and Howard, I. P. (1978). A Craik-O'Brien-Cornsweet illusion for visual depth. *Vision Research, 18*, 213–217.

Arnold, R. D. (1983). *Automated Stereo Perception.* Technical Report AIM-351, Artificial Intelligence Laboratory, Stanford University.

Baker, H. H. (1982). *Depth from Edge and Intensity Based Stereo.* Technical Report AIM-347, Artificial Intelligence Laboratory, Stanford University.

Baker, H. H. (1989). Building surfaces of evolution: the weaving wall. *International Journal of Computer Vision, 3(1)*.

Baker, H. H. and Bolles, R. C. (1989). Generalizing epipolar-plane image analysis on the spatiotemporal surface. *International Journal of Computer Vision, 3(1)*.

Ballard, D. H. (1987). Interpolation coding: a representation for numbers in neural models. *Biological Cybernetics, 57*, 389–402.

Barnard, S. T. (1986). A stochastic approach to stereo vision. In *Fifth National Conference on Artificial Intelligence (AAAI-86)*, pages 676–680, Morgan Kaufmann Publishers, Philadelphia, Pennsylvania.

Barnard, S. T. (1989). Stochastic stereo matching over scale. *International Journal of Computer Vision, 3(1)*.

Barnard, S. T. and Fischler, M. A. (1982). Computational stereo. *Computing Surveys, 14(4)*, 553–572.

Barrow, H. G. and Tenenbaum, J. M. (1978). Recovering intrinsic scene characteristics from images. In Hanson, A. R. and Riseman, E. M., editors, *Computer Vision Systems*, pages 3–26, Academic Press, New York, New York.

Barrow, H. G. and Tenenbaum, J. M. (1981). Interpreting line drawings as three-dimensional surfaces. *Artificial Intelligence, 17*, 75–116.

Bertero, M., Poggio, T., and Torre, V. (1987). *Ill-posed problems in early vision*. A. I. Memo 924, Massachusetts Institute of Technology.

Bierman, G. J. (1977). *Factorization Methods for Discrete Sequential Estimation*. Academic Press, New York, New York.

Blake, A. and Zisserman, A. (1986a). Invariant surface reconstruction using weak continuity constraints. In *IEEE Computer Society Conference on Computer Vision and Pattern Recognition (CVPR'86)*, pages 62–68, IEEE Computer Society Press, Miami Beach, Florida.

Blake, A. and Zisserman, A. (1986b). Some properties of weak continuity constraints and the GNC algorithm. In *IEEE Computer Society Conference on Computer Vision and Pattern Recognition (CVPR'86)*, pages 656–661, IEEE Computer Society Press, Miami Beach, Florida.

Blake, A. and Zisserman, A. (1987). *Visual Reconstruction*. MIT Press, Cambridge, Massachusetts.

Bolles, R. C. and Baker, H. H. (1985). Epipolar-plane image analysis: a technique for analysing motion sequences. In *Third International Symposium of Robotics Research*, pages 192–199, Gouvieux, France.

Bolles, R. C., Baker, H. H., and Marimont, D. H. (1987). Epipolar-plane image analysis: an approach to determining structure from motion. *International Journal of Computer Vision, 1*, 7–55.

Boult, T. E. (1986). *Information Based Complexity in Non-Linear Equations and Computer Vision*. Ph.D. thesis, Columbia University.

Bracewell, R. N. (1978). *The Fourier Transform and its Applications*. McGraw-Hill, New York, New York, 2nd edition.

Brooks, R. A., Greiner, R., and Binford, T. O. (1979). The ACRONYM model-based vision system. In *Sixth International Joint Conference on Artificial Intelligence (IJCAI-79)*, pages 105–113, Tokyo, Japan.

Burt, P. J. and Adelson, E. H. (1983). The Laplacian pyramid as a compact image code. *IEEE Transactions on Communications, COM-31(4)*, 532–540.

Canny, J. (1986). A computational approach to edge detection. *IEEE Transactions on Pattern Analysis and Machine Intelligence, PAMI-8(6)*, 679–698.

Chen, L. and Boult, T. E. (1988). An integrated approach to stereo matching, surface reconstruction and depth segmentation using consistent smoothness assumptions. In *Image Understanding Workshop*, pages 166–176, Morgan Kaufmann Publishers, Cambridge, Massachusetts.

Choi, D. J. (1987). Solving the depth interpolation problem on a fine grained, mesh- and tree-connected SIMD machine. In *Image Understanding Workshop*, pages 639–643, Morgan Kaufmann Publishers, Los Angeles, California.

Christ, J. P. (1987). *Shape Estimation and Object Recognition Using Spatial Probability Distributions*. Ph.D. thesis, Carnegie Mellon University.

Craven, P. and Wahba, G. (1979). Smoothing noisy data with spline functions: estimating the correct degree of smoothing by the method of generalized cross-validation. *Numerische Mathematik, 31*, 377–403.

Crowley, J. L. and Stern, R. M. (1982). *Fast Computation of the Difference of Low-Pass Transform*. Technical Report CMU-RI-TR-82-18, The Robotics Institute, Carnegie Mellon University.

Dev, P. (1974). *Segmentation Processes in Visual Perception: A Cooperative Neural Model*. COINS Technical Report 74C-5, University of Massachusetts at Amherst.

Drumheller, M. and Poggio, T. (1986). On parallel stereo. In *IEEE International Conference on Robotics and Automation*, pages 1439–1448, IEEE Computer Society Press, San Francisco, California.

Duda, R. O. and Hart, P. E. (1973). *Pattern Classification and Scene Analysis*. J. Wiley, New York, New York.

Durbin, R. and Willshaw, D. (1987). An analogue approach to the traveling salesman problem using an elastic net method. *Nature, 326*, 689–691.

Durbin, R., Szeliski, R., and Yuille, A. (1989). An analysis of the elastic net approach to the travelling salesman problem. In *Snowbird Neural Networks Meeting*, Snowbird, Utah.

Durrant-Whyte, H. F. (1987). Consistent integration and propagation of disparate sensor observations. *International Journal of Robotics Research, 6(3)*, 3–24.

Elfes, A. and Matthies, L. (1987). Sensor integration for robot navigation: combining sonar and stereo range data in a grid-based representation. In *IEEE Conference on Decision and Control*, IEEE Computer Society Press.

Faugeras, O. D. and Hebert, M. (1987). The representation, recognition and positioning of 3-D shapes from range data. In Kanade, T., editor, *Three-Dimensional Machine Vision*, pages 301–353, Kluwer Academic Publishers, Boston, Massachusetts.

Faugeras, O. D., Ayache, N., and Faverjon, B. (1986). Building visual maps by combining noisy stereo measurements. In *IEEE International Conference on Robotics and Automation*, pages 1433–1438, IEEE Computer Society Press, San Francisco, California.

Foley, T. A. (1987). Weighted bicubic spline interpolation to rapidly varying data. *ACM Transactions on Graphics, 6(1),* 1–18.

Fournier, A., Fussel, D., and Carpenter, L. (1982). Computer rendering of stochastic models. *Communications of the ACM, 25(6),* 371–384.

Gamble, E. and Poggio, T. (1987). *Visual integration and detection of discontinuities: the key role of intensity edges.* A. I. Memo 970, Artificial Intelligence Laboratory, Massachusetts Institute of Technology.

Gelb, A., editor. (1974). *Applied Optimal Estimation.* MIT Press, Cambridge, Massachusetts.

Geman, S. and Geman, D. (1984). Stochastic relaxation, Gibbs distribution, and the Bayesian restoration of images. *IEEE Transactions on Pattern Analysis and Machine Intelligence, PAMI-6(6),* 721–741.

Geman, S. and McClure, D. E. (1987). Statistical methods for tomographic image reconstruction. In *46th Session of the International Statistical Institute*, Bulletin of the ISI, vol. 52.

Gremban, K. D., Thorpe, C. E., and Kanade, T. (1988). Geometric camera calibration using systems of linear equations. In *IEEE International Conference on Robotics and Automation*, pages 562–567, IEEE Computer Society Press, Philadelphia, Pennsylvania.

Grimson, W. E. L. (1981). *From Images to Surfaces: a Computational Study of the Human Early Visual System.* MIT Press, Cambridge, Massachusetts.

Grimson, W. E. L. (1983). An implementation of a computational theory of visual surface interpolation. *Computer Vision, Graphics, and Image Processing, 22,* 39–69.

Hackbusch, W. (1985). *Multigrid Methods and Applications.* Springer-Verlag, Berlin.

Hackbusch, W. and Trottenberg, U., editors. (1982). *Multigrid Methods*, Springer-Verlag, Berlin, Heidelberg, New York.

Harris, J. G. (1987). A new approach to surface reconstruction: the coupled depth/slope model. In *First International Conference on Computer Vision (ICCV'87)*, pages 277–283, IEEE Computer Society Press, London, England.

Hebert, M. and Kanade, T. (1988). 3-D vision for outdoor navigation by an autonomous vehicle. In *Image Understanding Workshop*, pages 365–382, Morgan Kaufmann Publishers, Cambridge, Massachusetts.

Hebert, M., Kanade, T., and Kweon, I. (1988). *3-D Vision Techniques for Autonomous Vehicles.* Technical Report CMU-RI-TR-88-12, The Robotics

Institute, Carnegie Mellon University.

Heeger, D. J. (1986). Depth and flow from motion energy. In *Fifth National Conference on Artificial Intelligence (AAAI-86)*, pages 657–661, Morgan Kaufmann Publishers, Philadelphia, Pennsylvania.

Heeger, D. J. (1987). Optical flow from spatiotemporal filters. In *First International Conference on Computer Vision (ICCV'87)*, pages 181–190, IEEE Computer Society Press, London, England.

Henderson, R. L., Miller, W. J., and Grosch, C. B. (1979). Automated stereo reconstruction of man-made targets. *SPIE, 186(Digital Processing of Aerial Images)*, 240–248.

Hildreth, E. C. (1982). The integration of motion information along contours. In *IEEE Workshop on Computer Vision*, pages 83–91, IEEE Computer Society Press, Rindge, New Hampshire.

Hinton, G. E. (1977). *Relaxation and its Role in Vision*. Ph.D. thesis, University of Edinburgh.

Hinton, G. E. and Sejnowski, T. J. (1983). Optimal perceptual inference. In *IEEE Computer Society Conference on Computer Vision and Pattern Recognition (CVPR'83)*, pages 448–453, IEEE Computer Society Press, Washington, D. C.

Hoff, W. and Ahuja, N. (1986). Surfaces from stereo. In *Eighth International Conference on Pattern Recognition (ICPR'86)*, pages 516–518, IEEE Computer Society Press, Paris, France.

Hopfield, J. J. (1982). Neural networks and physical systems with emergent collective computational abilities. *Proceedings of the National Academy of Sciences U.S.A., 79*, 2554–2558.

Horn, B. K. P. (1977). Understanding image intensities. *Artificial Intelligence, 8*, 201–231.

Horn, B. K. P. and Brooks, M. J. (1986). The variational approach to shape from shading. *Computer Vision, Graphics, and Image Processing, 33*, 174–208.

Horn, B. K. P. and Schunck, B. G. (1981). Determining optical flow. *Artificial Intelligence, 17*, 185–203.

Hueckel, M. H. (1971). An operator which locates edges in digitized pictures. *Journal of the Association for Computing Machinery, 18(1)*, 113–125.

Hung, Y., Cooper, D. B., and Cernushi-Frias, B. (1988). Bayesian estimation of 3-D surfaces from a sequence of images. In *IEEE International Conference on Robotics and Automation*, pages 906–911, IEEE Computer Society Press, Philadelphia, Pennsylvania.

Ikeuchi, K. and Horn, B. K. P. (1981). Numerical shape from shading and occluding boundaries. *Artificial Intelligence, 17*, 141–184.

Julesz, B. (1971). *Foundations of Cyclopean Perception*. Chicago University Press, Chicago, Illinois.

Kanade, T. (1981). Recovery of the three-dimensional shape of an object from a single view. *Artificial Intelligence, 17*, 409–460.

Kass, M., Witkin, A., and Terzopoulos, D. (1988). Snakes: active contour models. *International Journal of Computer Vision, 1(4)*, 321–331.

Kass, M. H. (1984). *Computing Stereo Correspondence*. Master's thesis, Massachusetts Institute of Technology.

Kimeldorf, G. and Wahba, G. (1970). A correspondence between Bayesian estimation on stochastic processes and smoothing by splines. *The Annals of Mathematical Statistics, 41(2)*, 495–502.

Kirkpatrick, S., Gelatt, C. D. J., and Vecchi, M. P. (1983). Optimization by simulated annealing. *Science, 220*, 671–680.

Koch, C., Marroquin, J., and Yuille, A. (1986). Analog "neuronal" networks in early vision. *Proceedings of the National Academy of Sciences U.S.A., 83*, 4263–4267.

Konrad, J. and Dubois, E. (1988). Multigrid bayesian estimation of image motion fields using stochastic relaxation. In *Second International Conference on Computer Vision (ICCV'88)*, pages 354–362, IEEE Computer Society Press, Tampa, Florida.

Leclerc, Y. G. (1989). Constructing simple stable descriptions for image partitioning. *International Journal of Computer Vision, 3(1)*, 75–104.

Leclerc, Y. G. and Zucker, S. W. (1987). The local structure of image discontinuities in one dimension. *IEEE Transactions on Pattern Analysis and Machine Intelligence, PAMI-9(3)*, 341–355.

Lehky, S. R. and Sejnowski, T. J. (1988). Network model of shape-from-shading: neural function arises from both receptive and projective fields. *Nature, 333*, 452–454.

Lewis, J. P. (1987). Generalized stochastic subdivision. *ACM Transactions on Graphics, 6(3)*, 167–190.

Lowe, D. G. (1985). *Perceptual Organization and Visual Recognition*. Kluwer Academic Publishers, Boston, Massachusetts.

Lucas, B. D. (1984). *Generalized Image Matching by the Method of Differences*. Ph.D. thesis, Carnegie Mellon University.

Mallat, S. G. (1987). Scale change versus scale space representation. In *First International Conference on Computer Vision (ICCV'87)*, pages 592–596, IEEE Computer Society Press, London, England.

Mandelbrot, B. B. (1982). *The Fractal Geometry of Nature*. W. H. Freeman, San Francisco, California.

Marr, D. (1978). Representing visual information. In Hanson, A. R. and Riseman, E. M., editors, *Computer Vision Systems*, pages 61–80, Academic Press, New York, New York.

Marr, D. (1982). *Vision: A Computational Investigation into the Human Representation and Processing of Visual Information*. W. H. Freeman, San Francisco, California.

Marr, D. and Hildreth, E. (1980). Theory of edge detection. *Proceedings of the Royal Society of London, B 207*, 187–217.

Marr, D. and Poggio, T. (1976). Cooperative computation of stereo disparity. *Science, 194*, 283–287.

Marr, D. C. and Poggio, T. (1979). A computational theory of human stereo vision. *Proceedings of the Royal Society of London, B 204*, 301–328.

Marroquin, J. L. (1984). *Surface Reconstruction Preserving Discontinuities.* A. I. Memo 792, Artificial Intelligence Laboratory, Massachusetts Institute of Technology.

Marroquin, J. L. (1985). *Probabilistic Solution of Inverse Problems.* Ph.D. thesis, Massachusetts Institute of Technology.

Matthies, L. H. and Shafer, S. A. (1987). Error modeling in stereo navigation. *IEEE Journal of Robotics and Automation, RA-3(3)*, 239–248.

Matthies, L. H., Szeliski, R., and Kanade, T. (1987). *Kalman Filter-based Algorithms for Estimating Depth from Image Sequences.* Technical Report CMU-CS-87-185, Computer Science Department, Carnegie Mellon University.

Matthies, L. H., Szeliski, R., and Kanade, T. (1989). Kalman filter-based algorithms for estimating depth from image sequences. *International Journal of Computer Vision*, .

Maybeck, P. S. (1979). *Stochastic Models, Estimation, and Control.* Volume 1, Academic Press, New York, New York.

Maybeck, P. S. (1982). *Stochastic Models, Estimation, and Control.* Volume 2, Academic Press, New York, New York.

McDermott, D. (1980). *Spatial Inferences with Ground, Metric Formulas on Simple Objects.* Research Report 173, Department of Computer Science, Yale University.

Mead, C. A. and Mahowald, M. A. (1988). A silicon model of early visual processing. *Neural Networks, 1*, 91–97.

Metropolis, N., Rosenbluth, A. W., Rosenbluth, M. N., Teller, A. H., and Teller, E. (1953). Equations of state calculations by fast computing machines. *Journal of Chemical Physics, 21*, 1087–1091.

Moravec, H. P. (1988). Sensor fusion in certainty grids for mobile robots. *AI Magazine, 9(2)*, 61–74.

Nalwa, V. (1986). On detecting edges. *IEEE Transactions on Pattern Analysis and Machine Intelligence, PAMI-8(6)*, 699–714.

Ohta, Y. and Kanade, T. (1985). Stereo by intra- and inter-scanline search using dynamic programming. *IEEE Transactions on Pattern Analysis and Machine Intelligence, PAMI-7(2)*, 139–154.

Pentland, A. P. (1984). Fractal-based description of natural scenes. *IEEE Transactions on Pattern Analysis and Machine Intelligence, PAMI-6(6)*, 661–674.

Pentland, A. P. (1986). Perceptual organization and the representation of natural form. *Artificial Intelligence, 28(3),* 293–331.

Poggio, T. *et al..* (1988). The MIT vision machine. In *Image Understanding Workshop,* pages 177–198, Morgan Kaufmann Publishers, Boston, Massachusetts.

Poggio, T. and Torre, V. (1984). Ill-posed problems and regularization analysis in early vision. In *Image Understanding Workshop,* pages 257–263, Science Applications International Corporation, New Orleans, Louisiana.

Poggio, T., Torre, V., and Koch, C. (1985a). Computational vision and regularization theory. *Nature, 317(6035),* 314–319.

Poggio, T., Voorhees, H., and Yuille, A. (1985b). *A Regularized Solution to Edge Detection.* A. I. Memo 833, Artificial Intelligence Laboratory, Massachusetts Institute of Technology.

Prazdny, K. (1985). Detection of binocular disparities. *Biological Cybernetics,* 52, 93–99.

Press, W. *et al..* (1986). *Numerical Recipes: The Art of Scientific Computing.* Cambridge University Press, Cambridge, England.

Quam, L. H. (1984). Hierarchical warp stereo. In *Image Understanding Workshop,* pages 149–155, Science Applications International Corporation, New Orleans, Louisiana. Also available as Technical Note No. 402, SRI International, December, 1986.

Rensink, R. A. (1986). *On the Visual Discrimination of Self-Similar Random Textures.* Master's thesis, The University of British Columbia.

Rives, P., Breuil, E., and Espiau, B. (1986). Recursive estimation of 3D features using optical flow and camera motion. In *Conference on Intelligent Autonomous Systems,* pages 522–532, Elsevier Science Publishers. (also appeared in 1987 IEEE International Conference on Robotics and Automation).

Roberts, L. G. (1965). Machine perception of three-dimensional solids. In Tippett *et al.,* editors, *Optical and Electro-Optical Information Processing,* chapter 9, pages 159–197, MIT Press, Cambridge, Massachusetts.

Rosenfeld, A. (1980). Quadtrees and pyramids for pattern recognition and image processing. In *Fifth International Conference on Pattern Recognition (ICPR'80),* pages 802–809, IEEE Computer Society Press, Miami Beach, Florida.

Rosenfeld, A., editor. (1984). *Multiresolution Image Processing and Analysis,* Springer-Verlag, New York, New York.

Rosenfeld, A. and Kak, A. C. (1976). *Digital Picture Processing.* Academic Press, New York, New York.

Rosenfeld, A., Hummel, R. A., and Zucker, S. W. (1976). Scene labeling by relaxation operations. *IEEE Transactions on Systems, Man, and Cybernetics,* SMC-6, 420–433.

Rumelhart, D. E., Hinton, G. E., and Williams, R. J. (1986). Learning internal representations by error propagation. In Rumelhart, D. E., McClelland, J. L., and the PDP research group, editors, *Parallel distributed processing: Explorations in the microstructure of cognition*, Bradford Books, Cambridge, Massachusetts.

Shafer, S. A. (1988). *Automation and Calibration for Robot Vision Systems*. Technical Report CMU-CS-88-147, Computer Science Department, Carnegie Mellon University.

Shafer, S. A. and Kanade, T. (1983). *The Theory of Straight Homogeneous Generalized Cylinders and A Taxonomy of Generalized Cylinders*. Technical Report CMU-CS-83-105, Computer Science Department, Carnegie Mellon University.

Sivilotti, M. A., Mahowald, M. A., and Mead, C. A. (1987). Real-time visual computations using analog CMOS processing arrays. In Losleben, P., editor, *Advanced Research in VLSI: Proceedings of the 1987 Stanford Conference*, pages 295–312, MIT Press, Cambridge, Massachusetts.

Stewart, W. K. (1987). A non-deterministic approach to 3-D modeling underwater. In *Fifth International Symposium on Unmanned Untethered Submersible Technology*, pages 283–309, University of New Hampshire Marine Systems Engineering Laboratory.

Szeliski, R. (1986). *Cooperative Algorithms for Solving Random-Dot Stereograms*. Technical Report CMU-CS-86-133, Computer Science Department, Carnegie Mellon University.

Szeliski, R. (1987). Regularization uses fractal priors. In *Sixth National Conference on Artificial Intelligence (AAAI-87)*, pages 749–754, Morgan Kaufmann Publishers, Seattle, Washington.

Szeliski, R. (1988a). Estimating motion from sparse range data without correspondence. In *Second International Conference on Computer Vision (ICCV'88)*, pages 207–216, IEEE Computer Society Press, Tampa, Florida.

Szeliski, R. (1988b). Some experiments in calibrating the Sony CCD camera. Internal report, IUS group, Carnegie Mellon University.

Szeliski, R. (1989). Fast surface interpolation using hierarchical basis functions. In *IEEE Computer Society Conference on Computer Vision and Pattern Recognition (CVPR'89)*, IEEE Computer Society Press, San Diego, California.

Szeliski, R. and Hinton, G. (1985). Solving random-dot stereograms using the heat equation. In *IEEE Computer Society Conference on Computer Vision and Pattern Recognition (CVPR'85)*, pages 284–288, IEEE Computer Society Press, San Francisco, California.

Szeliski, R. and Terzopoulos, D. (1989a). From splines to fractals. *Computer Graphics (SIGGRAPH'87)*, 23(4).

Szeliski, R. and Terzopoulos, D. (1989b). Parallel multigrid algorithms and applications to computer vision. In *Fourth Copper Mountain Conference on Multigrid Methods*, (Preprints), Copper Mountain, Colorado.

Terzopoulos, D. (1983). Multilevel computational processes for visual surface reconstruction. *Computer Vision, Graphics, and Image Processing*, 24, 52–96.

Terzopoulos, D. (1984). *Multiresolution Computation of Visible-Surface Representations*. Ph.D. thesis, Massachusetts Institute of Technology.

Terzopoulos, D. (1985). Concurrent multilevel relaxation. In Baumann, L. S., editor, *Image Understanding Workshop*, pages 156–162, Science Applications International Corporation, Miami Beach, Florida.

Terzopoulos, D. (1986a). Image analysis using multigrid relaxation methods. *IEEE Transactions on Pattern Analysis and Machine Intelligence*, PAMI-8(2), 129–139.

Terzopoulos, D. (1986b). Regularization of inverse visual problems involving discontinuities. *IEEE Transactions on Pattern Analysis and Machine Intelligence*, PAMI-8(4), 413–424.

Terzopoulos, D. (1987). Matching deformable models to images: direct and iterative solutions. In *Topical Meeting on Machine Vision*, pages 164–167, Optical Society of America, Washington, D. C.

Terzopoulos, D. (1988). The computation of visible-surface representations. *IEEE Transactions on Pattern Analysis and Machine Intelligence*, PAMI-10(4), 417–438.

Terzopoulos, D., Witkin, A., and Kass, M. (1987). Symmetry-seeking models and 3D object reconstruction. *International Journal of Computer Vision*, 1(3), 211–221.

Tikhonov, A. N. and Arsenin, V. Y. (1977). *Solutions of Ill-Posed Problems*. V. H. Winston, Washington, D. C.

Tsai, R. Y. and Huang, T. S. (1984). Uniqueness and estimation of three-dimensional motion parameters of rigid objects with curved surfaces. *IEEE Transactions on Pattern Analysis and Machine Intelligence*, PAMI-6(1), 13–27.

Ullman, S. (1979). *The Interpretation of Visual Motion*. MIT Press, Cambridge, Massachusetts.

Van Essen, D. C. and Maunsell, J. H. R. (1983). Hierarchical organization and functional streams in the visual cortex. *Trends in Neuroscience*, 6, 370–375.

Voss, R. F. (1985). Random fractal forgeries. In Earnshaw, R. A., editor, *Fundamental Algorithms for Computer Graphics*, Springer-Verlag, Berlin.

Wahba, G. (1983). Bayesian "confidence intervals" for the cross-validated smoothing spline. *Journal of the Royal Statistical Society*, B 45(1), 133–150.

Waltz, D. L. (1975). Understanding line drawings of scenes with shadows.

In Winston, P., editor, *The Psychology of Computer Vision*, McGraw-Hill, New York, New York.

Webb, J. A. and Aggarwal, J. K. (1981). Visually interpreting the motion of objects in space. *Computer, 14(8)*, 40–46.

Wilson, K. G. (1979). Problems in physics with many scales of length. *Scientific American, 241(2)*, 158–179.

Witkin, A., Terzopoulos, D., and Kass, M. (1987). Signal matching through scale space. *International Journal of Computer Vision, 1*, 133–144.

Witkin, A. P. (1981). Recovering surface shape and orientation from texture. *Artificial Intelligence, 17*, 17–45.

Witkin, A. P. (1983). Scale-space filtering. In *Eighth International Joint Conference on Artificial Intelligence (IJCAI-83)*, pages 1019–1022, Morgan Kaufmann Publishers.

Woodham, R. J. (1981). Analysing images of curved surfaces. *Artificial Intelligence, 17*, 117–140.

Yserentant, H. (1986). On the multi-level splitting of finite element spaces. *Numerische Mathematik, 49*, 379–412.

Zucker, S. W. (1986). Early orientation selection: inferring trace, tangent, and curvature fields. In *Eighth International Conference on Pattern Recognition (ICPR'86)*, pages 294–302, IEEE Computer Society Press, Paris, France.

Appendix A

Finite element implementation

In this Appendix, we present the discrete implementation of the energy equations used in Section 2.2. Our implementation is based on previous work described by Terzopoulos (1984), Marroquin (1984), Blake and Zisserman (1987), and Harris (1987). The energy equations which we develop are formulated in terms of a number of discrete fields which form the basic data structures used in the surface interpolation algorithm:

$$
\begin{array}{ll}
u_{i,j} & \text{surface depth (free variables)} \\
d_{i,j} & \text{depth constraints} \\
c_{i,j} & \text{depth constraint weights } (\sigma^{-2}) \\
l_{i,j}, m_{i,j} & \text{depth discontinuities } [0,1] \\
n_{i,j} & \text{orientation discontinuities } [0,1]
\end{array}
$$

The line variables $l_{i,j}$ and $m_{i,j}$ are located on a dual grid, as shown in Figure 2.16. The crease variables $n_{i,j}$ are coincident with the depth value nodes.

To describe the discrete version of the smoothness constraint, we define a number of *implicit* fields which are not actually computed by the algorithm but serve only as a notational shorthand. First, we define the finite differences

$$u_{i,j}^x = u_{i+1,j} - u_{i,j}$$
$$u_{i,j}^y = u_{i,j+1} - u_{i,j}$$
$$u_{i,j}^{xx} = u_{i+1,j}^x - u_{i,j}^x = u_{i+2,j} - 2u_{i+1,j} + u_{i,j}$$
$$u_{i,j}^{xy} = u_{i,j+1}^x - u_{i,j}^x = u_{i+1,j}^y - u_{i,j}^y = u_{i+1,j+i} - u_{i+1,j} - u_{i,j+1} + u_{i,j}$$
$$u_{i,j}^{yy} = u_{i,j+1}^y - u_{i,j}^y = u_{i,j+2} - 2u_{i,j+1} + u_{i,j}.$$

Next, we define the continuity strengths

$$\beta_{i,j}^x = (1 - l_{i,j})$$
$$\beta_{i,j}^y = (1 - m_{i,j})$$
$$\beta_{i,j}^{xx} = (1 - l_{i,j})(1 - l_{i+1,j})(1 - n_{i+1,j})$$

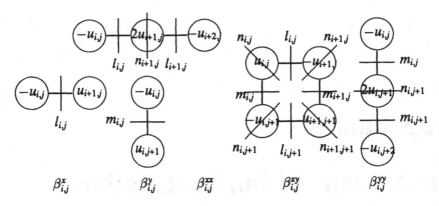

$$\beta^x_{i,j} \qquad \beta^y_{i,j} \qquad \beta^{xx}_{i,j} \qquad \beta^{xy}_{i,j} \qquad \beta^{yy}_{i,j}$$

Figure A.1: Continuity strengths and computational molecules

$$\beta^{xy}_{i,j} = (1 - l_{i,j})(1 - l_{i,j+1})(1 - m_{i,j})(1 - m_{i+1,j})(1 - n_{i,j}n_{i+1,j+1})(1 - n_{i+1,j}n_{i,j+1})$$
$$\beta^{yy}_{i,j} = (1 - m_{i,j})(1 - m_{i,j+1})(1 - n_{i,j+1}).$$

These continuity strengths determine which of the *molecules* defined by the $u^x_{i,j}, \ldots, u^{yy}_{i,j}$ finite differences are active (we use phantom line variables around the edges of the grid to take care of boundary conditions).

The interactions between the depth field, the line and crease variables, and the continuity strengths are shown in Figure A.1. Intuitively, the line variables disable any molecules that are cut when the line variable is turned on. The crease variables disable the $u^{xx}_{i,j}$ or $u^{yy}_{i,j}$ molecules whose centers are coincident with the crease, and disable the $u^{xy}_{i,j}$ molecules if two opposite corners are creased. Note that instead of splicing in a small membrane element at a crease location (as is done by Terzopoulos (1984)), we simply disable the hinge (second order) element centered on the crease. Creases in the surface thus do not have a tendency to flatten the surface locally.

The prior energy (weak smoothness constraint) is constructed from these elements

$$E_p(\mathbf{u}) = \frac{1}{2} \sum_{(i,j)} \begin{aligned} & w_0 h^2 \left[(u_{i,j})^2\right] + w_1 \left[\beta^x_{i,j}(u^x_{i,j})^2 + \beta^y_{i,j}(u^y_{i,j})^2\right] \\ & + w_2 h^{-2} \left[\beta^{xx}_{i,j}(u^{xx}_{i,j})^2 + 2\beta^{xy}_{i,j}(u^{xy}_{i,j})^2 + \beta^{yy}_{i,j}(u^{yy}_{i,j})^2\right] \end{aligned} \qquad (A.1)$$

where $h = \Delta x = \Delta y$ is the grid spacing. The weighting functions w_m, which come from the general controlled-continuity constraint (2.6), define the order of the interpolator. In terms of the notation used by Terzopoulos (1986b) which is shown in (2.5), we have

$$\{w_0, w_1, w_2\} = \{0, \rho(x,y)[1 - \tau(x,y)], \rho(x,y)\tau(x,y)\}$$

(in our implementation, we use the line and crease variables to represent the discontinuities rather than spatially varying w_m). In terms of the notation used by Blake and Zisserman (1987), we have

$$\{w_0, w_1, w_2\} = \{0, \lambda^2, \mu^4\}.$$

Blake and Zisserman (1987) claim that the λ and μ parameters are more natural, since they correspond to the *characteristic lengths* or scales of the smoothing filters implied by regularization (Appendix B). Our own preference for using the w_m parameterization comes from its usefulness when designing interpolators with fractional degrees of continuity (Section 4.2).

The equation for the data compatibility constraint is the same as we presented in Section 2.2, namely

$$E_d(\mathbf{u}, \mathbf{d}) = \frac{1}{2} \sum_{(i,j)} c_{i,j} (u_{i,j} - d_{i,j})^2 \qquad (A.2)$$

with $c_{i,j} = \sigma_{i,j}^{-2}$ being the inverse variance of the measurement $d_{i,j}$. If we wish to add penalty terms for the lines and creases (because we are performing joint optimization of the surface and its discontinuities), we can use the same equation as used by Blake and Zisserman (1987)

$$E_c(\mathbf{l}, \mathbf{m}, \mathbf{n}) = \sum_{(i,j)} \alpha_l l_{i,j} + \alpha_m m_{i,j} + \alpha_n n_{i,j}. \qquad (A.3)$$

More complicated energies which try to enforce continuity in the lines could also be used (Geman and Geman 1984).

The stiffness matrix $\mathbf{A}$ and the weighted data vector $\mathbf{b}$ can be obtained from the discrete energy equation

$$E(\mathbf{u}) = E_p(\mathbf{u}) + E_d(\mathbf{u}, \mathbf{d})$$

by setting

$$a_{(i,j),(k,l)} = \frac{\partial^2}{\partial u_{i,j} \partial u_{k,l}} E(\mathbf{u}) \Big|_{\mathbf{u}=0} \qquad (A.4)$$

and

$$b_{i,j} = -\frac{\partial}{\partial u_{i,j}} E(\mathbf{u}) \Big|_{\mathbf{u}=0}. \qquad (A.5)$$

In the case where $E(\mathbf{u})$ is not quadratic, we can evaluate the above expressions at the current estimate $\mathbf{u}$, and solve $\mathbf{A}\Delta\mathbf{u} = \mathbf{b}$.

To derive the energy equations at a coarse level from the fine level equations, we have two choices. We can use the interpolation equations that map from the coarse to the fine level to derive the new equations; alternatively, we can compute the coarse level data constraints $d_{i,j}$ and $c_{i,j}$ and the line and crease variables $l_{i,j}$,

$m_{i,j}$ and $n_{i,j}$ from the fine level, and then use the usual finite-element analysis to obtain the discrete energy equations.

Using the first approach, we choose an interpolation matrix $\mathbf{F}_c^f$ to map from the coarse to the fine level

$$\mathbf{u}^f = \mathbf{F}_c^f \mathbf{u}^c.$$

The fine level quadratic equation can thus be re-written as

$$
\begin{aligned}
E(\mathbf{u}^f) &= \mathbf{u}^{f^T} \mathbf{A}^f \mathbf{u}^f - \mathbf{u}^{f^T} \mathbf{b}^f + k \\
&= (\mathbf{F}_c^f \mathbf{u}^c)^T \mathbf{A}^f (\mathbf{F}_c^f \mathbf{u}^c) - \mathbf{u}^{c^T} \mathbf{F}_c^{f^T} \mathbf{b}^f + k \\
&= \mathbf{u}^{c^T} \mathbf{A}^c \mathbf{u}^c - \mathbf{u}^{c^T} \mathbf{b}^c + k
\end{aligned}
$$

where

$$\mathbf{A}^c = \mathbf{F}_c^{f^T} \mathbf{A}^f \mathbf{F}_c^f \quad \text{and} \quad \mathbf{b}^c = \mathbf{F}_c^{f^T} \mathbf{b}^f.$$

The problem with this approach is that the neighborhoods implementing the smoothness constraint get progressively larger and lose their simple structure. The smoothing behavior, which can be determined by taking the Fourier transform of an entry from $\mathbf{A}$ (see Appendix B) also becomes worse (compared to that obtained using the original finite-element equations).

For this reason, we have chosen to use the second approach. The depth value constraints for the coarse level are obtained from the fine level constraints by weighted averaging

$$c_{i,j}^c = \sum_{k=0}^{1} \sum_{l=0}^{1} c_{2i+k,2j+l}^f \tag{A.6}$$

and

$$d_{i,j}^c = \frac{1}{c_{i,j}^c} \sum_{k=0}^{1} \sum_{l=0}^{1} c_{2i+k,2j+l}^f d_{2i+k,2j+l}^f. \tag{A.7}$$

This corresponds to assuming a block interpolation function. Using more sophisticated interpolation functions will give better coarse level constraints, but will also introduce off-diagonal terms into the data constraints (Section 5.1).

The line and crease variables are similarly obtained from the fine level equations using the logical OR function,

$$l_{i,j}^c = \bigvee_{k=0}^{1} \bigvee_{l=0}^{1} l_{2i+k,2j+l}^f$$

$$m_{i,j}^c = \bigvee_{k=0}^{1} \bigvee_{l=0}^{1} m_{2i+k,2j+l}^f$$

$$n_{i,j}^c = \bigvee_{k=0}^{1} \bigvee_{l=0}^{1} n_{2i+k,2j+l}^f.$$

Again, more sophisticated combination rules could be used, but they have not been implemented for reasons of simplicity. Once all of the new fields $d_{i,j}, \ldots, n_{i,j}$ have been derived for the coarse level, we can compute the discrete smoothness and data compatibility equations as before.

Appendix B

Fourier analysis

By taking a Fourier transform of the function $u(\mathbf{x})$ and expressing the energy equations in the frequency domain, we can analyze the spectral properties of our Bayesian models and the convergence properties of our estimation algorithms. In Section 4.1, we used Fourier analysis to compute the spectral characteristics of the prior model. In this Appendix, we will analyze the filtering behavior of regularization and the spectral characteristics of the posterior model. We will also examine the convergence properties of gradient descent and the frequency response characteristics of our discrete implementation. We will then use these results to determine the convergence properties of multigrid relaxation, and to demonstrate its advantages both for deterministic MAP estimation and for stochastic sampling.

B.1 Filtering behavior of regularization

As we showed in Section 4.1, the energy of the smoothness constraint (prior model) used in regularization can be expressed in the frequency domain as

$$E_p(U) = \frac{1}{2} \int |H_p(\mathbf{f})|^2 |U(\mathbf{f})|^2 d\mathbf{f} \tag{B.1}$$

where

$$|H_p(\mathbf{f})|^2 = \sum_{m=0}^{p} w_m |2\pi \mathbf{f}|^{2m}.$$

Calculating the energy associated with the smoothness constraint is thus equivalent to passing the signal $u(\mathbf{x})$ through a filter whose impulse response is $h_p(\mathbf{x})$, and then taking the usual (signal squared) energy measure

$$E_p(\mathbf{u}) = \frac{1}{2} \int |h_p(\mathbf{x}) * u(\mathbf{x})|^2 d\mathbf{x}. \tag{B.2}$$

Similarly, if we assume that the measured data $d(\mathbf{x})$ is dense and has uniform weighting σ^{-2}, we can derive from (2.7) the data compatibility term expressed in the Fourier domain

$$E_d(U) = \frac{1}{2} \int |H_d(f)|^2 |U(f) - D(f)|^2 df \qquad (B.3)$$

where

$$|H_d(f)|^2 = \sigma^{-2}.$$

Thus, the overall energy function, written in the frequency domain, is

$$E(U) = \frac{1}{2} \int |H(f)|^2 |U(f)|^2 - 2|H_d(f)|^2 |U(f)||D(f)| + |H_d(f)|^2 |D(f)|^2 df \qquad (B.4)$$

where

$$|H(f)|^2 = |H_p(f)|^2 + |H_d(f)|^2.$$

The minimum energy solution for $U(f)$ can be calculated as

$$U^*(f) = \frac{|H_d(f)|^2}{|H(f)|^2} D(f). \qquad (B.5)$$

In the spatial domain, this is equivalent to filtering the data $d(\mathbf{x})$ with a shift-invariant filter $h_s(\mathbf{x})$,

$$u^*(\mathbf{x}) = h_s(\mathbf{x}) * d(\mathbf{x}) \qquad (B.6)$$

where

$$h_s(\mathbf{x}) = \mathcal{F}^{-1} \{H_s(f)\} = \mathcal{F}^{-1} \left\{ \frac{|H_d(f)|^2}{|H(f)|^2} \right\}. \qquad (B.7)$$

This equivalence between regularization and convolution (filtering) has previously been noted by Poggio and Torre (1984) and Terzopoulos (1986b).

For the membrane surface interpolator we have

$$|H_d(f)|^2 = \sigma^{-2} \quad \text{and} \quad |H_p(f)|^2 = |2\pi f|^2$$

with a resulting smoothing filter

$$H_s(f) = \frac{1}{1 + \sigma^2 |2\pi f|^2}. \qquad (B.8)$$

Similarly, for the thin plate surface interpolator we have

$$H_s(f) = \frac{1}{1 + \sigma^2 |2\pi f|^4}. \qquad (B.9)$$

The shape of the frequency response for these two models is qualitatively similar to that of a Gaussian filter. The exact equations for the one- and two-dimensional impulse responses corresponding to these filters are given in (Poggio *et al.* 1985b).

B.2   Fourier analysis of the posterior distribution

In the previous section, we saw how for the case of dense uniform data, the results of using regularization are equivalent to convolution with a shift-invariant filter $h_s(x)$. If we are using Bayesian modeling instead of regularization, however, the posterior estimate is characterized by a probability distribution rather than just the optimal estimate u^*. To characterize the posterior distribution $p(u|d)$, we can look at the *residual* signal

$$\mathbf{v} \equiv \mathbf{u} - \mathbf{u}^*,$$

i.e., the difference between the signal and the optimal estimate. Substituting $U^*(f) + V(f)$ for $U(f)$ in (B.4), we obtain

$$E(U^*+V) = \frac{1}{2} \int |H(f)|^2 |U^*(f)+V(f)|^2 - 2|H_d(f)|^2|U^*(f)+V(f)||D(f)|+|H_d(f)|^2|D(f)|^2 df.$$

Using (B.5), this simplifies to

$$E(V) = \frac{1}{2} \int |H(f)|^2 |V(f)|^2 df + k. \tag{B.10}$$

Following the same reasoning as in Section 4.1, we see that $v(x)$ is pink (correlated) Gaussian noise with a power spectrum

$$P_v(f) = |H(f)|^{-2}. \tag{B.11}$$

The estimated signal $u(x)$ is thus the combination of the filtered data $u^*(x)$ and the pink noise $v(x)$

$$u(x) = u^*(x) + v(x) = h_s(x) * d(x) + h^{-1}(x) * w(x) \tag{B.12}$$

where $w(x)$ is white Gaussian noise (Figure B.1). The Fourier transform of this estimate is

$$U(f) = H_s(f)D(f) + |H(f)|^{-1}W(f), \tag{B.13}$$

where $W(f)$ is the Fourier transform of white Gaussian noise, and is hence itself a white Gaussian signal. Taking expectations, we have

$$\langle U(f) \rangle = H_s(f)D(f) = U^*(f) \tag{B.14}$$

and

$$P_u(f) = \langle U^2(f) \rangle = |H_s(f)|^2|D(f)|^2 + |H(f)|^{-2}. \tag{B.15}$$

Thus, if we define the power spectrum to be the average power (signal squared) in a narrow frequency band we have the above result. If we define the power spectrum to be the variance in the Fourier transform, we have

$$P_u(f) = \text{Var}(U(f)) = \langle U^2(f) \rangle - \langle U(f) \rangle^2 = |H(f)|^{-2}. \tag{B.16}$$

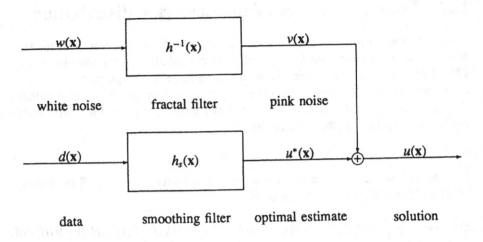

Figure B.1: Filter and noise model

B.3   Analysis of gradient descent

In this section we will perform a Fourier analysis of the gradient descent technique which we use to estimate the MAP solution, and of the stochastic gradient descent technique (Gibbs Sampler) which we use to generate random samples. For the moment, we will confine our analysis to the continuous-space case. The next section will discuss the finite element (discrete-space) implementation.

Gradient descent techniques (which include most of the relaxation algorithms which we examined in Section 2.2) iteratively move the system state towards its optimal solution by taking small steps in the direction of the energy gradient[1]. To analyze the convergence of the algorithm towards the final solution, we can either compute the gradient in the spatial or the frequency domain (this is possible because the $u(\mathbf{x})$ and its transform $U(\mathbf{f})$ are linear transforms of each other). In this section, we will take the latter approach since it results in a simpler exposition. We will also confine our analysis to the discrete time case. The continuous time case (which corresponds to analog systems such as those studied by Koch *et al.* (1986)) can be analyzed using differential equations instead of difference equations.

The gradient of the energy equation (B.4) with respect to $U(\mathbf{f})$ is given by

$$\nabla E = \frac{\partial E(U)}{\partial U(\mathbf{f})} = |H(\mathbf{f})|^2 U(\mathbf{f}) - |H_d(\mathbf{f})|^2 D(\mathbf{f}).$$

---

[1]For stochastic gradient descent, we modify this rule by adding a small random disturbance at each step.

For the discrete time system, the updating equation is

$$U(\mathbf{f}, t+1) = U(\mathbf{f}, t) - \alpha \nabla E = [1 - \alpha |H(\mathbf{f})|^2] U(\mathbf{f}, t) + \alpha |H_d(\mathbf{f})|^2 D(\mathbf{f})$$

where α controls the step size of the gradient descent. The solution of this recurrence relation is

$$U(\mathbf{f}, t) = U(\mathbf{f}, 0)[1 - \alpha |H(\mathbf{f})|^2]^t + \frac{|H_d(\mathbf{f})|^2}{|H(\mathbf{f})|^2} D(\mathbf{f}) \left(1 - [1 - \alpha |H(\mathbf{f})|^2]^t\right). \quad \text{(B.17)}$$

Note that the geometric series converges only if $\alpha |H(\mathbf{f})|^2 < 2$, which is not in general true for the controlled-continuity constraint. However, as we will see in the next section, the discrete space system has strictly bounded responses for $|H(\mathbf{f})|^2$.

To analyze the convergence of stochastic relaxation (the Gibbs Sampler), we will ignore the component of the solution that is converging towards the optimal solution, and only look at the residual $v(\mathbf{x})$. In this case, we use the stochastic updating equation

$$V(\mathbf{f}, t+1) = V(\mathbf{f}, t) - \alpha[\nabla E + \gamma W(\mathbf{f}, t)] = [1 - \alpha |H(\mathbf{f})|^2] V(\mathbf{f}, t) - \alpha \gamma W(\mathbf{f}, t),$$

where $W(\mathbf{f}, t)$ is the Fourier transform of the white noise field $w(\mathbf{x})$ that is added to the solution at time t, and γ is the magnitude of this noise. The recurrence relation for the power spectrum of $v(\mathbf{x}, t)$ can be computed from the above equation as

$$P(\mathbf{f}, t+1) = [1 - \alpha |H(\mathbf{f})|^2]^2 P(\mathbf{f}, t) + \alpha^2 \gamma^2.$$

The solution of this recurrence relation is

$$P(\mathbf{f}, t) = P(\mathbf{f}, 0)[1 - \alpha |H(\mathbf{f})|^2]^{2t} + \frac{\alpha^2 \gamma^2}{1 - [1 - \alpha |H(\mathbf{f})|^2]^2} \left(1 - [1 - \alpha |H(\mathbf{f})|^2]^{2t}\right).$$

$$\text{(B.18)}$$

Thus, if we set $\gamma = \sqrt{2/\alpha}$, the steady state power spectrum is

$$P(\mathbf{f}, \infty) = |H(\mathbf{f})|^{-2}[1 - \frac{\alpha}{2} |H(\mathbf{f})|^2]^{-1} \quad \text{(B.19)}$$

which approaches the true spectrum (B.11) as $\alpha \to 0$. The discrepancy between the power spectrum of stochastic Jacobi relaxation and that of the Markov Random Field is due to the synchronous updating of the stochastic system (the Gibbs Sampler uses asynchronous updating).

B.4   Finite element solution

The finite-element energy approximations for the membrane and the thin plate are derived by Terzopoulos (1984) by considering particular local spline elements, showing that they are acceptable using a patch test, and then calculating the energy in terms of the nodal variables. This approach is also useful for the computation of energy terms near discontinuities, as we saw in Appendix A. An alternative approach is to consider the system as a discrete approximation to the filtering system analyzed in Section B.1.

Following this latter approach, we desire a prior filter whose frequency response is

$$|H_p(f)|^2 = \sum_{m=0}^{p} w_m |2\pi f|^{2m}.$$

This can be approximated by a digital filter whose frequency response is

$$H'_p(f) = \sum_{m=0}^{p} w_m h^{-2m} \left(2n - \sum_i 2 \cos 2\pi f_i h \right)^m \qquad \text{(B.20)}$$

where $h = |\Delta x| = |\Delta y|$ is the grid size, and n is the dimensionality of the x domain[2]. The impulse response of the individual filters $l_m(x)$ are the finite difference approximations to the Laplacian, i.e., $l_0(x) = \delta(x)$,

$$l_1(x) = h^{-2} \begin{cases} 2n & \text{if } x = 0 \\ -1 & \text{if } |x| = 1 \\ 0 & \text{otherwise} \end{cases},$$

and

$$l_m(x) = \underbrace{l_1(x) * \cdots * l_1(x)}_{m \text{ times}}.$$

The integral equation that describes the energy functional

$$E_p(u) = \frac{1}{2} \int |h_p(x) * u(x)|^2 dx$$

can now be replaced by the summation

$$E_p(u) = \frac{1}{2} \sum |h'_p(x) * u(x)|^2 h^n. \qquad \text{(B.21)}$$

We therefore have to modify the above definitions of $l_m(x)$ by scaling up[3] by h^n.

---

[2] The small ω approximation for $\cos \omega$ is $1 - \omega^2/2! + \omega^4/4! - \ldots$

[3] The scaling by h is required to make the multigrid version algorithm have consistent energy approximations, since the magnitude of the energy determines the variance of the stochastic $u(x)$ signal.

For two dimensions, the l_1 and l_2 operators correspond to the membrane and thin plate neighborhoods[4]. The shape of these neighborhoods is

$$
\begin{bmatrix} & -1 & \\ -1 & 4 & -1 \\ & -1 & \end{bmatrix}
\quad \text{and} \quad
h^{-2}
\begin{bmatrix}
 & & 1 & & \\
 & 2 & -8 & 2 & \\
1 & -8 & 20 & -8 & 1 \\
 & 2 & -8 & 2 & \\
 & & 1 & &
\end{bmatrix}.
$$

Note that these neighborhoods are not the only ones possible for the finite-element approximation, just the simplest to implement on a digital computer. For example, we could use the filter

$$
\begin{bmatrix}
-1 & -1 & -1 \\
-1 & 8 & -1 \\
-1 & -1 & -1
\end{bmatrix}
$$

for the membrane interpolator, since it has the correct small f approximation.

B.5 Fourier analysis of multigrid relaxation

The Fourier domain tools developed in the previous sections can be used to analyze the convergence rates of multigrid relaxation. In this section, we combine the Jacobi gradient descent technique examined in Section B.3 with the discrete filters obtained in Section B.4 to perform this analysis. As a starting point we set $U(\mathbf{f}, 0) = 0$, and use a data compatibility filter $|H_d(\mathbf{f})|^2 = \sigma^{-2}$. We will define the *effective filter response* $F(\mathbf{f}, t)$ as the ratio between the Fourier transform of the solution and the Fourier transform of the data

$$
F(\mathbf{f}, t) \equiv \frac{U(\mathbf{f}, t)}{D(\mathbf{f})} = \frac{|H_d(\mathbf{f})|^2}{|H(\mathbf{f})|^2} \left(1 - [1 - \alpha |H(\mathbf{f})|^2]^t \right). \tag{B.22}
$$

For a (two dimensional) thin plate interpolator on a discrete grid of size h, we have

$$
|H_d(\mathbf{f})|^2 = h^2 \sigma^{-2}
$$

and

$$
|H(\mathbf{f})|^2 = h^{-2}(4 - 2\cos 2\pi f_x h - 2\cos 2\pi f_y h)^2 + h^2 \sigma^{-2}.
$$

To ensure that the Jacobi iteration converges (without ringing) we will make the conservative assumption that $\alpha |H(\mathbf{f})|^2 \le 1$. Since $|H(\mathbf{f})|^2$ attains a maximum of $64h^{-2} + h^2 \sigma^{-2}$ at (π, π), we set $\alpha^{-1} = 64h^{-2} + h^2 \sigma^{-2}$.

---

[4]These neighborhoods correspond to the rows of the $\mathbf{A}_p$ matrix.

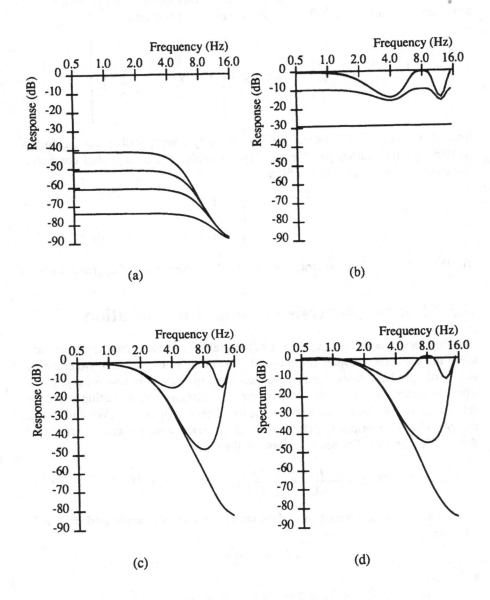

Figure B.2: Effective frequency response of multigrid relaxation

(a) fine level only (b) coarse level only (c) multigrid (d) spectral density

The frequency response $F((f_x, 0), t)$ with $\sigma = 0.005$ and $h = 1/32$ (33×33 points) is shown in Figure B.2a after 10, 25, 50 and 100 iterations. Note how slowly the low-frequency response approaches the equilibrium value. On a coarser grid ($h = 1/8$), the low-frequency convergence—shown in Figure B.2b after 1, 5, 25 and 50 iterations—is much faster, but the high frequencies are not annihilated. Multigrid relaxation is shown in Figure B.2c, with 25 iterations at $h = 1/8$, 25 iterations at $h = 1/16$, and a final 25 iterations at $h = 1/32$.

A similar analysis can be applied to the spectral density of the random field. Recall from (B.18) that

$$P(\mathbf{f}, t) = |H(\mathbf{f})|^{-2}[1 - \frac{\alpha}{2}|H(\mathbf{f})|^2]^{-1} \left(1 - [1 - \alpha|H(\mathbf{f})|^2]^{2t}\right).$$

The plots of the spectral density as a function of time are shown in Figure B.2d. Note that even though the Jacobi iteration does not converge to the true spectral density $|H(\mathbf{f})|^{-2}$, the error is not very large.

While the Fourier analysis on which these results are based is valid only for the spatially isotropic dense data case, these results can still be used to explain the improved convergence obtained with multigrid relaxation. First, we note that the main advantage of multigrid relaxation is that it approximates more quickly the *low* frequency filtering (smoothing) behavior of the system, while ignoring the exact *high* frequency response. Second, the use of the power spectrum for the stochastic signal gives us a measure of how quickly the system is approaching thermal equilibrium. This is because a multivariate Gaussian distribution (which results from a quadratic energy function) can be completely characterized by its current mean and variance. For the spatially isotropic case just examined, the effective filter response and power spectrum capture these statistics in a succinct fashion. The convergence rate results which we have presented in this Appendix can also be applied to number of the algorithms studied in this book, including the multiresolution fractal generation algorithms (Section 4.2) and the Monte Carlo variance estimator (Section 6.2).

Appendix C

Analysis of optical flow computation

In this Appendix, we analyze the performance of the simple correlation-based flow estimator introduced in Section 5.3. This analysis is an extension of the one-dimensional analysis presented in (Matthies *et al.* 1987). The flow estimator which we examine selects at each pixel the disparity that minimizes the SSD error measure

$$e_t(\mathbf{d}; \mathbf{x}) = \int w(\lambda)[f_1(\mathbf{x} + \mathbf{d} + \lambda) - f_0(\mathbf{x} + \lambda)]^2 \, d\lambda, \qquad (C.1)$$

where $f_0(\mathbf{x})$ and $f_1(\mathbf{x})$ are the two successive image frames, and $w(\mathbf{x})$ is a symmetric, non-negative weighting function. To perform our analysis, we will assume that the two image frames are generated from an underlying true intensity image, $f(\mathbf{x})$, to which uncorrelated (white) Gaussian noise with variance σ_n^2 has been added:

$$f_0(\mathbf{x}) = f(\mathbf{x}) + n_0(\mathbf{x}),$$
$$f_1(\mathbf{x} + \tilde{\mathbf{d}}) = f(\mathbf{x}) + n_1(\mathbf{x}).$$

Using this model, we can rewrite the error measure as[1]

$$e(\mathbf{d}; \mathbf{x}) = \int w(\lambda)[f(\mathbf{x} + \mathbf{d} - \tilde{\mathbf{d}} + \lambda) - f(\mathbf{x} + \lambda) + n_1(\mathbf{x} + \lambda) - n_0(\mathbf{x} + \lambda)]^2 \, d\lambda.$$

If $\mathbf{d} \simeq \tilde{\mathbf{d}}$, we can use a Taylor series expansion to obtain

$$e(\mathbf{d}; \mathbf{x}) = \int w(\lambda)[\nabla f(\mathbf{x} + \lambda) \cdot (\mathbf{d} - \tilde{\mathbf{d}}) + n_1(\mathbf{x} + \lambda) - n_0(\mathbf{x} + \lambda)]^2 \, d\lambda$$

$$= (\mathbf{d} - \tilde{\mathbf{d}})^T \mathbf{A}(\mathbf{d} - \tilde{\mathbf{d}}) + 2(\mathbf{b}_1(\mathbf{x}) - \mathbf{b}_0(\mathbf{x}))^T(\mathbf{d} - \tilde{\mathbf{d}}) + c \qquad (C.2)$$

---

[1] This equation is actually incorrect, since it should contain $n_1(\mathbf{x} + \mathbf{d} - \tilde{\mathbf{d}} + \lambda)$ instead of $n_1(\mathbf{x} + \lambda)$. The effect of including the correct term is to add small random terms involving integrals of $w(\lambda)$, $\nabla w(\lambda)$, $\nabla f(\mathbf{x} + \lambda)$, $\nabla \nabla f(\mathbf{x} + \lambda)$ and $n_1(\mathbf{x})$ to the quadratic coefficient $\mathbf{A}(\mathbf{x})$, $\mathbf{b}_1(\mathbf{x})$ and $c(\mathbf{x})$ that are derived below. This intentional omission has been made to simplify the presentation.

where

$$A = \int w(\lambda)\nabla f(x + \lambda)\nabla^T f(x + \lambda)\, d\lambda,$$

$$b_k = \int w(\lambda)\nabla f(x + \lambda)n_k(x + \lambda)\, d\lambda,$$

$$c = \int w(\lambda)[n_1(x + \lambda) - n_0(x + \lambda)]^2\, d\lambda.$$

The four coefficients A, b_0, b_1 and c define the shape of the error surface $e(d; x)$. The first coefficient, A, is related to the average roughness or slope of the intensity surface, and determines the confidence given to the disparity estimate (see below). If $\nabla f(x)$ is relatively constant, this matrix has rank 1, and we have the classical aperture problem. The second and third coefficients, b_0 and b_1, are independent zero mean Gaussian random variables that determine the difference between the displacement estimate $\hat{d}$ and the true displacement $\bar{d}$, (i.e., the error in flow estimator). The fourth coefficient, c, is a chi-squared distributed random variable with mean $(2\sigma_n^2 \int w(\lambda)\, d\lambda)$ that defines the height of the error surface at $d = \bar{d}$. We can thus use c to obtain a local estimate of the image noise. High values of c can also be used to flag areas—such as occluded areas—where the flow computation model is inaccurate.

To estimate the disparity at a point x given the error surface $e(d; x)$, we find the displacement vector $\hat{d}$ such that

$$e(\hat{d}; x) = \min_{d} e(d; x).$$

By differentiating the quadratic[2] equation (C.2), we see that this minimum will be at

$$\hat{d} = \bar{d} + A^{-1}(b_0 - b_1). \tag{C.3}$$

To calculate the variance in this estimate, we must first calculate the variance in b_k,

$$\text{Var}(b_k) = \langle b_k b_k^T \rangle = \sigma_n^2 \int w^2(\lambda)\nabla f(x + \lambda)\nabla^T f(x + \lambda)\, d\lambda.$$

If we set $w(x) = 1$ on some finite interval and zero elsewhere, this ensures that $w^2(\lambda) = w(\lambda)$, and the variance reduces to $\sigma_n^2 A$. If an arbitrary window function $w(x)$ is used, we can scale the variance estimate by $(\int w^2(\lambda)\, d\lambda)/(\int w(\lambda)\, d\lambda)$, which is valid if ∇f is relatively constant with respect to the window size.

Combining the above results, we conclude that

$$\text{Var}(\hat{d}) = \langle (\hat{d} - \bar{d})(\hat{d} - \bar{d})^T \rangle = A^{-1} \langle (b_0 - b_1)(b_0 - b_1)^T \rangle A^{-1} = 2\sigma_n^2 A^{-1} \tag{C.4}$$

[2]The true equation (when higher order Taylor series terms are included) is a polynomial series in $(d - \bar{d})$ with random coefficients of decreasing variance. This explains the rough nature of the $e(d; x)$ observed in practice.

since b_0 and b_1 are uncorrelated. The covariance matrix for the displacement estimate can thus be determined from the quadratic form matrix fitted to the error surface. Empirical studies performed on real images with a one-dimensional version of this flow estimator have shown a good fit between the estimated and actual variances (Matthies *et al.* 1987). The approach used in this Appendix can also be used to calculate the spatial or temporal correlations between flow estimates by performing a two-dimensional generalization of the derivation given by Matthies *et al.* (1987).

Appendix D

Analysis of parameter estimation

In this Appendix, we will develop our Bayesian estimation theory in terms of multivariate Gaussian distributions and derive some of the results that were needed in Sections 4.2, 6.3, and 6.4. We will use the notation $x \sim N(m, P)$ to denote that x is a multivariate normal variable with mean m and covariance P (Gelb 1974). The probability density function for x can be written as

$$p(x) = |2\pi P|^{-1/2} \exp\left(-\frac{1}{2}(x - m)^T P^{-1}(x - m)\right) \qquad (D.1)$$

where $|P|$ is the determinant of the matrix P (we use $|2\pi P|^{-1/2}$ instead of $(2\pi)^{-n/2}|P|^{-1/2}$ for notational succinctness).

D.1   Computing marginal distributions

The first result which we wish to establish concerns computing marginal distributions from a joint distribution. Consider the distribution

$$p\left(\begin{bmatrix} x \\ y \end{bmatrix}\right) \propto \exp -\frac{1}{2} \begin{bmatrix} x - x_0 \\ y - y_0 \end{bmatrix}^T \begin{bmatrix} A & B \\ B^T & C \end{bmatrix} \begin{bmatrix} x - x_0 \\ y - y_0 \end{bmatrix}^T.$$

This probability distribution attains a maximum w.r.t. x (corresponding to the minimum of the quadratic form) for

$$\hat{x} = x_0 - A^{-1}B(y - y_0).$$

We can diagonalize the probability distribution by substituting for x_0 in the original equation

$$p\left(\begin{bmatrix} x \\ y \end{bmatrix}\right) \propto \exp -\frac{1}{2} \begin{bmatrix} x - \hat{x} - A^{-1}B(y - y_0) \\ y - y_0 \end{bmatrix}^T \begin{bmatrix} A & B \\ B^T & C \end{bmatrix} \begin{bmatrix} x - \hat{x} - A^{-1}B(y - y_0) \\ y - y_0 \end{bmatrix}$$

$$= \exp -\frac{1}{2}\begin{bmatrix} \mathbf{x} - \hat{\mathbf{x}} \\ \mathbf{y} - \mathbf{y}_0 \end{bmatrix}^T \begin{bmatrix} \mathbf{I} & -\mathbf{A}^{-1}\mathbf{B} \\ 0 & \mathbf{I} \end{bmatrix}^T \begin{bmatrix} \mathbf{A} & \mathbf{B} \\ \mathbf{B}^T & \mathbf{C} \end{bmatrix} \begin{bmatrix} \mathbf{I} & -\mathbf{A}^{-1}\mathbf{B} \\ 0 & \mathbf{I} \end{bmatrix} \begin{bmatrix} \mathbf{x} - \hat{\mathbf{x}} \\ \mathbf{y} - \mathbf{y}_0 \end{bmatrix}$$

$$= \exp -\frac{1}{2}\begin{bmatrix} \mathbf{x} - \hat{\mathbf{x}} \\ \mathbf{y} - \mathbf{y}_0 \end{bmatrix}^T \begin{bmatrix} \mathbf{A} & 0 \\ 0 & \mathbf{C} - \mathbf{B}^T\mathbf{A}^{-1}\mathbf{B} \end{bmatrix} \begin{bmatrix} \mathbf{x} - \hat{\mathbf{x}} \\ \mathbf{y} - \mathbf{y}_0 \end{bmatrix}$$

Because of this diagonal structure, we can marginalize w.r.t. $\mathbf{x}$ to obtain

$$p(\mathbf{y}) \propto \exp -\frac{1}{2}(\mathbf{y} - \mathbf{y}_0)^T(\mathbf{C} - \mathbf{B}^T\mathbf{A}^{-1}\mathbf{B})(\mathbf{y} - \mathbf{y}_0).$$

However, evaluating $\mathbf{A}^{-1}$ may be expensive, especially if $\mathbf{A}$ is large and sparse.

The alternative to performing this evaluation is to compute $\hat{\mathbf{x}}$ in terms of the current value of $\mathbf{y}$ (this usually involves solving a sparse set of linear equations), and to substitute this value into the probability density function

$$p\left(\begin{bmatrix} \hat{\mathbf{x}} \\ \mathbf{y} \end{bmatrix}\right) \propto \exp -\frac{1}{2}\begin{bmatrix} -\mathbf{A}^{-1}\mathbf{B}(\mathbf{y} - \mathbf{y}_0) \\ \mathbf{y} - \mathbf{y}_0 \end{bmatrix}^T \begin{bmatrix} \mathbf{A} & \mathbf{B} \\ \mathbf{B}^T & \mathbf{C} \end{bmatrix} \begin{bmatrix} -\mathbf{A}^{-1}\mathbf{B}(\mathbf{y} - \mathbf{y}_0) \\ \mathbf{y} - \mathbf{y}_0 \end{bmatrix}$$

$$= \exp -\frac{1}{2}\begin{bmatrix} \mathbf{x} - \hat{\mathbf{x}} \\ \mathbf{y} - \mathbf{y}_0 \end{bmatrix}^T \begin{bmatrix} -\mathbf{A}^{-1}\mathbf{B} \\ \mathbf{I} \end{bmatrix}^T \begin{bmatrix} \mathbf{A} & \mathbf{B} \\ \mathbf{B}^T & \mathbf{C} \end{bmatrix} \begin{bmatrix} -\mathbf{A}^{-1}\mathbf{B} \\ \mathbf{I} \end{bmatrix} \begin{bmatrix} \mathbf{x} - \hat{\mathbf{x}} \\ \mathbf{y} - \mathbf{y}_0 \end{bmatrix}$$

$$= \exp -\frac{1}{2}(\mathbf{y} - \mathbf{y}_0)^T(\mathbf{C} - \mathbf{B}^T\mathbf{A}^{-1}\mathbf{B})(\mathbf{y} - \mathbf{y}_0).$$

Thus, we see that marginalizing a multivariate Gaussian distribution with respect to some of the variables is equivalent to substituting the minimum energy solution for those variables into the original equation (this result is used in Section 4.2).

D.2 Bayesian estimation equations

In the remainder of this Appendix, we will develop the equations for the various conditional and marginal distributions which we use in our Bayesian estimation framework. The notation which we use is derived from the Kalman filtering literature (Gelb 1974). In this Appendix, we assume that all of our observations come from a static surface characterized by the state vector $\mathbf{u}$; in Section 7.1, we extend our estimator to include a dynamic system model.

We start by modeling the prior distribution as a correlated Gaussian

$$\mathbf{u} \sim N(\hat{\mathbf{u}}_0, \mathbf{P}_0). \tag{D.2}$$

For our sensor model, we use a linear system with additive Gaussian noise

$$\mathbf{d} = \mathbf{H}\mathbf{u} + \mathbf{r}, \quad \text{with} \quad \mathbf{r} \sim N(0, \mathbf{R}). \tag{D.3}$$

The measurement matrix $\mathbf{H}$ encodes the sparse sampling which converts from the dense depth map $\mathbf{u}$ to the sparse set of depth values $\mathbf{d}$. This rectangular matrix usually contains 1's where the nodal variables coincide with the data points, and 0's elsewhere. If the data points do not lie on the grid (e.g., if we have sub-pixel positioning accuracy), then the $\mathbf{H}$ matrix can be derived from the local interpolation function and will contain non-integer values (see Section 5.1). The $\mathbf{R}$ matrix encodes the covariance of the measurement noise process, and is usually diagonal with $r_{ii} = \sigma_i^2$ for uncorrelated sensor noise.

From the sensor model (D.3), we can derive the distribution of the data $\mathbf{d}$ conditioned on the initial state $\mathbf{u}$ as

$$\mathbf{d}|\mathbf{u} \sim N(\mathbf{Hu}, \mathbf{R}). \tag{D.4}$$

Similarly, we can derive the marginal distribution of the data by integrating over all possible initial states to obtain

$$\mathbf{d} \sim N(\mathbf{H\hat{u}_0}, \mathbf{HP_0H}^T + \mathbf{R}) \tag{D.5}$$

(this result can also be derived using the formula for sums of Gaussian random variables).

The posterior estimate $\mathbf{u}_1$ after the first set of measurements can be derived from (D.2) and (D.4) using Bayes' Rule. This estimate is a multivariate Gaussian

$$\mathbf{u}_1 \sim N(\mathbf{\hat{u}_1}, \mathbf{P}_1) \tag{D.6}$$

with a mean

$$\mathbf{\hat{u}_1} = (\mathbf{P}_0^{-1} + \mathbf{H}^T\mathbf{R}^{-1}\mathbf{H})^{-1}(\mathbf{P}_0^{-1}\mathbf{\hat{u}_0} + \mathbf{H}^T\mathbf{R}^{-1}\mathbf{d}) \tag{D.7}$$

and a covariance

$$\mathbf{P}_1 = (\mathbf{P}_0^{-1} + \mathbf{H}^T\mathbf{R}^{-1}\mathbf{H})^{-1} \tag{D.8}$$

(see (Gelb 1974) or (Maybeck 1979) for derivations).

The mean estimate $\mathbf{\hat{u}_1}$, which is also the MAP estimate, corresponds to the minimum energy solution of the spline approximation (2.15) if we make the following correspondences

$$\mathbf{\hat{u}_0} = 0, \quad \mathbf{P}_0^{-1} = \mathbf{A_p}, \quad \mathbf{H}^T\mathbf{R}^{-1}\mathbf{H} = \mathbf{A_d}, \quad \text{and} \quad \mathbf{Hd} = \mathbf{\bar{d}},$$

where $\mathbf{\bar{d}}$ is the zero-padded observation vector. The advantage of using an explicit measurement equation (D.3) is that unlike $\mathbf{A_d}$, $\mathbf{R}$ is not singular and can easily model correlated sensor noise. The measurement matrix $\mathbf{H}$ also allows us to model a wider variety of sensors, since measurements need not be coincident with grid locations. Finally, the prior state estimate $\mathbf{\hat{u}_0}$ can be used to encode any prior knowledge which we have about the surface (e.g., from a digital terrain map in navigation applications).

The new state and state covariance estimates $\hat{u}_1$ and P_1 capture all of the information that is available from the prior model and the first observation. These values could then be substituted into (D.2) and used to obtain posterior estimates $\hat{u}_2$ and P_2 from a second set of measurements. In general, the Bayesian model can be extended to include multiple measurements of the same surface using

$$d_k = H_k u + r_k, \quad \text{with} \quad r_k \sim N(0, R_k).$$

After n measurements have been processed, the posterior estimate has a mean value

$$\hat{u}_n = (P_n^{-1})^{-1} \sum_{k=1}^{n} H_k^T R_k^{-1} d_k$$

and a covariance

$$P_n = \left(P_0^{-1} + \sum_{k=1}^{n} H_k^T R_k^{-1} H_k \right)^{-1}.$$

These equations can be re-written into a recursive form using

$$A_k = P_0^{-1} + \sum_{j=1}^{k} H_j^T R_j^{-1} H_j = A_{k-1} + H_k^T R_k^{-1} H_k \qquad (D.9)$$

and

$$b_k = \sum_{j=1}^{k} H_j^T R_j^{-1} d_j = b_{k-1} + H_k^T R_k^{-1} d_k \qquad (D.10)$$

to accumulate the inverse covariance and cumulative weighted data vector. The A_k's and b_k's are the same as are used in spline interpolation of multiple data sets (A_k is sparse and banded). The optimal estimate at time k can be obtained by solving

$$A_k \hat{u}_k = b_k. \qquad (D.11)$$

D.3   Likelihood of observations

The parameter estimation techniques developed in Sections 6.3 and 6.4 are based on maximizing the likelihood of having observed the given data. This likelihood is described by (D.5), and can thus be written as

$$p(d) = |2\pi(HP_0 H^T + R)|^{-1/2} \exp -\frac{1}{2}(d - H\hat{u}_0)^T (HP_0 H^T + R)^{-1}(d - H\hat{u}_0). \quad (D.12)$$

This probability distribution, however, is difficult to evaluate since it involves computing the covariance matrix P_0, which is not sparse (or may not even exist).

To get around this problem, we will re-write this equation in terms of quantities such as $\hat{u}_0$, $\hat{u}_1$, P_0^{-1}, R^{-1} and P_1^{-1} which we know how to compute. In order to do this, we will need to introduce two Lemmas. The first is a matrix inversion Lemma (Maybeck 1979, p. 280)

$$(HPH^T + R)^{-1} = R^{-1} - R^{-1}H(H^TR^{-1}H + P^{-1})^{-1}H^TR^{-1} \qquad (D.13)$$

where H need not be a square matrix. The second Lemma is related to matrix determinants

$$|HPH^T + R|^{-1} = |R^{-1}||P^{-1}||H^TR^{-1}H + P^{-1}|^{-1} \qquad (D.14)$$

(Maybeck 1979, p. 280).

We can re-write the energy equation corresponding to the negative logarithm of (D.12) as

$$E(\mathbf{d}) = \frac{1}{2}\log|2\pi(HP_0H^T + R)| + \frac{1}{2}(\mathbf{d} - H\hat{u}_0)^T(HP_0H^T + R)^{-1}(\mathbf{d} - H\hat{u}_0)\}$$
$$= E_1(\mathbf{d}) + E_2(\mathbf{d})$$

Applying the (D.14) and (D.8) to the first term, we obtain

$$E_1(\mathbf{d}) = \frac{1}{2}\log|P_1^{-1}| - \frac{1}{2}\log|2\pi R^{-1}| - \frac{1}{2}\log|P_0^{-1}|. \qquad (D.15)$$

We can re-write the second term of this energy equation by applying (D.13), (D.7) and (D.8)

$$E_2(\mathbf{d}) = \frac{1}{2}(\mathbf{d} - H\hat{u}_0)^T[R^{-1} - R^{-1}H(P_0^{-1} + H^TR^{-1}H)^{-1}H^TR^{-1}](\mathbf{d} - H\hat{u}_0)$$

$$= \frac{1}{2}(\mathbf{d} - H\hat{u}_0)^TR^{-1}[(\mathbf{d} - H\hat{u}_0) - HP_1(H^TR^{-1}\mathbf{d} - H^TR^{-1}H\hat{u}_0)]$$

$$= \frac{1}{2}(\mathbf{d} - H\hat{u}_0)^TR^{-1}[(\mathbf{d} - H\hat{u}_0) - HP_1((P_1^{-1}\hat{u}_1 - P_0^{-1}\hat{u}_0) - (P_1^{-1} - P_0^{-1})\hat{u}_0)]$$

$$= \frac{1}{2}(\mathbf{d} - H\hat{u}_0)^TR^{-1}(\mathbf{d} - H\hat{u}_0) - \frac{1}{2}(\mathbf{d} - H\hat{u}_0)^TR^{-1}H(\hat{u}_1 - \hat{u}_0) \qquad (D.16)$$

$$= \frac{1}{2}(\mathbf{d} - H\hat{u}_0)^TR^{-1}(\mathbf{d} - H\hat{u}_1) \qquad (D.17)$$

$$= \frac{1}{2}(\mathbf{d} - H\hat{u}_1)^TR^{-1}(\mathbf{d} - H\hat{u}_1) + \frac{1}{2}(\mathbf{d} - H\hat{u}_1)^TR^{-1}H(\hat{u}_1 - \hat{u}_0) \qquad (D.18)$$

$$= \frac{1}{2}(\mathbf{d} - H\hat{u}_1)^TR^{-1}(\mathbf{d} - H\hat{u}_1) + \frac{1}{2}(\hat{u}_1 - \hat{u}_0)^TP_0^{-1}(\hat{u}_1 - \hat{u}_0). \qquad (D.19)$$

These alternate forms of the energy equation have different interpretations. The form given in (D.16) shows that if we measure the energy using the residual between the data points and the prior surface estimate, we must reduce the energy by a term proportional to how much the surface moves as it is updated. Similarly, in (D.18) we see that using the residual between the data and the posterior estimate underestimates the energy by the same amount. The form given in (D.17) uses a cross-product between the residual to the prior estimate and the residual to the posterior estimate. Finally, (D.19) is a symmetric form which adds the data compatibility energy computed with respect to the posterior estimate to the strain (prior model) energy associated with the difference in solutions.

Table of symbols

SYMBOL	MEANING
λ	regularization parameter
μ	mean value
σ^2	noise or estimate variance
$\mathbf{b}$	cumulative weighted data vector
$d(x, y)$	data signal
$d_{i,j}$	data point
$\mathbf{d}$	data vector (or disparity)
$\mathbf{f}$	frequency (or intensity)
h	finite element size
$l_{i,j}, \; m_{i,j}$	line variables
$n_{i,j}$	crease variables
$\mathbf{p}$	three dimensional point position
$\mathbf{q}$	system noise
$\mathbf{r}$	measurement noise
$u(x, y)$	solution or state
$u_{i,j}$	discrete form of $u(x, y)$
$\mathbf{u}$	state vector
$\mathbf{u}^*$	optimal solution
$\hat{\mathbf{u}}$	current estimate
$\mathbf{v}$	hierarchical representation
$\mathbf{x}$	spatial domain
$\mathbf{A}$	cumulative state information matrix
$\mathbf{A_p}$	prior information matrix
$\mathbf{A_d}$	data information matrix
$\mathbf{C}$	three dimensional point covariance matrix
$E_d(\mathbf{u}, \mathbf{d})$	data compatibility energy
$E_p(\mathbf{u})$	prior energy
$E(\mathbf{u})$	combined energy
$\mathbf{F}$	state transition matrix

H	measurement matrix
K	Kalman filter gain matrix
$N(\mu, \sigma^2)$	normal (Gaussian) distribution
P	state covariance matrix
Q	system noise covariance matrix
R	measurement noise covariance matrix
S	hierarchical basis set matrix
T	temperature (in Gibbs distribution)
Z	partition function (in Gibbs distribution)

Index